T0143265

CATCHING NATURE IN THE ACT

Catching **NATURE** *in the Act*

✲

Réaumur and the Practice of Natural History in the Eighteenth Century

✲

Mary Terrall

The University of Chicago Press
Chicago & London

MARY TERRALL is professor of history at the University of California, Los Angeles. She is the auhor of *The Man Who Flattened the Earth: Maupertuis and the Sciences in the Enlightenment*, also published by the University of Chicago Press.

✲

The University of Chicago Press, Chicago 60637
The University of Chicago Press, Ltd., London

Printed in the United States of America

✲

23 22 21 20 19 18 17 16 15 14 1 2 3 4 5

✲

ISBN-13: 978-0-226-08860-0 (cloth)
ISBN-13: 978-0-226-08874-7 (e-book)
DOI: 10.7208/chicago/9780226088747.001.0001

✲

Library of Congress Cataloging-in-Pubication Data

Terrall, Mary, author.
Catching nature in the act : Réaumur and the practice of natural history in the eighteenth century / Mary Terrall.
pages cm
Includes bibliographical references and index.
ISBN 978-0-226-08860-0 (cloth : alkaline paper)
ISBN 978-0-226-08874-7 (e-book)
1. Natural history—History—18th century. 2. Réaumur, René-Antoine Ferchault de, 1683–1757. 3. Naturalists—History—18th century. I. Title
QH15.T47 2014
508—dc23
2013016604

✲

♾ This paper meets the requirements of ANSI/NISO Z39.48-1992 (Permanence of Paper).

✲

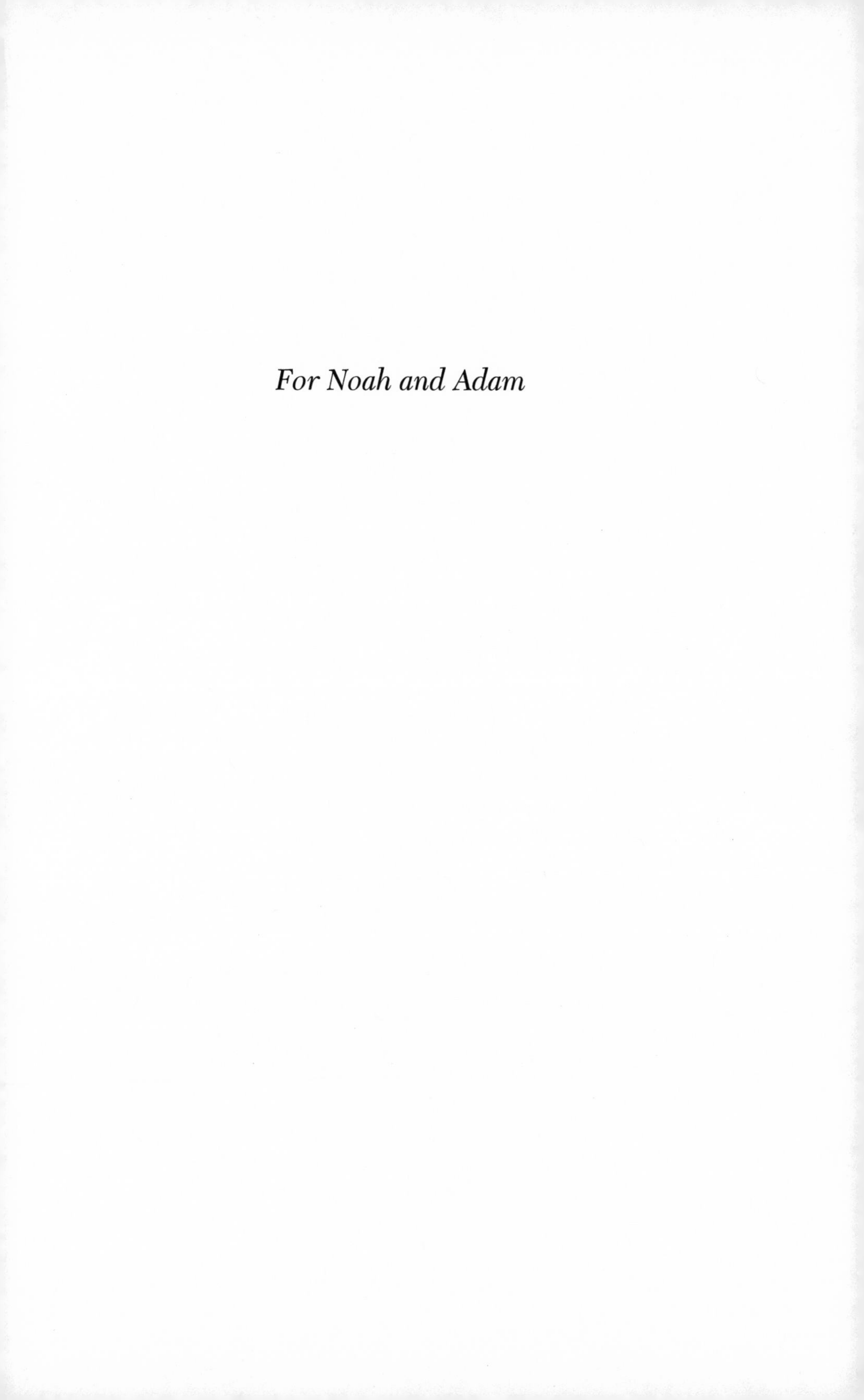

For Noah and Adam

CONTENTS

1

The Terrain of Natural History

An eighteenth-century reader familiar with contemporary natural history would recognize this passage by the Dutch naturalist and artist Pierre Lyonet as rather typical of the genre:

> The fact [of insect respiration] seems to me clearly demonstrated for many kinds of aquatic insects; I mean those that one often sees lifting the end of the tail to the surface of the water, and remaining there as if suspended [from the surface]. These tails are their organs of respiration, and they only hold them thus in the air in order to breathe. To convince yourself of this, you have only to cover the surface of the water where they are kept, with something that prevents them from lifting their tails to the surface. Right away you will see them rush about and search with extraordinary distress for some opening where they can insert this extremity of their body. If they do not find such an opening, very soon you see them go to the bottom and die, often in much less time than it would take to drown the most delicate terrestrial insect. . . . Anyone who might wish to do this experiment should be aware that not all aquatic insects that breathe through the hind end die equally quickly when they are prevented from getting to the air.[1]

This snippet of text exposes to view a peculiarity of insect life—the behavior of a particular kind of water bug, and the deduction that insects require air for respiration—alongside instructions for readers who might want to see such things for themselves. The naturalist-author assumes the participation of his readers, with such offhand phrases as "those that one often sees" or "you will see them rush about," and demonstrates almost casually how to devise an experiment to test a proposition and how to reason from the results. Lyonet could almost be talking to someone across a tub full of aquatic insects ("the water where they are kept"), giving his interlocutor the benefit of his long experience with such creatures. He knows how long it takes to drown terrestrial insects as well as different sorts of aquatic ones not enumerated here. He has evidently lived with these insects, not only watching their movements, but interfering with them—depriving them of air and timing their death throes.

In Lyonet's day, natural history was many things to many people—diversion, obsession, medically or economically useful knowledge, evidence for God's providence and wisdom, or even the foundation of all natural knowledge. Everything from the realms of animal, vegetable, and mineral became grist for the mill of its investigations. This science was so extensive and multifaceted, and practiced by such a variety of people around the globe, that it could hardly be considered a discipline, nor did its practitioners necessarily share institutions, training, or theoretical predilections. They were, however, unified by their dedication to observing, cultivating, chasing, collecting, experimenting, dissecting, preserving, drawing, and describing all manner of creatures (as well as plants and fossils and rocks). Pursuing these activities with varying degrees of intensity, they formed elaborate clusters and networks of exchange, collaboration, and debate.[2]

An exploration of these exchanges and disputes will expose the dynamics of the production of knowledge about the living world and the way that different kinds of people made natural history an intimate part of their daily lives. Natural history was at once diffuse, with its subject matter everywhere, and intensely focused, as its practitioners amassed minute and seemingly inconsequential details of structure or function. Working from the letters, drawings, and printed material that circulated through a dispersed community of naturalists, I attempt in this book to capture the motivations and interests that drove people to invest enormous time and effort into investigation of the intricacies of insects, worms, birds, and such. By picking out some interconnected individuals, studies of particular species, and large theoretical issues, I assemble a collage of scientific practices operating in different registers and on different scales across Europe and around the globe. Along the way, I examine the place of natural history relative to other sciences in the francophone world of the early Enlightenment. For the most part, I leave aside taxonomic questions, because these were not central concerns of the men and women whose observations and collections fill the pages of this book. In leaving aside taxonomy, I am, of course, leaving aside botany, where classification and nomenclature were central matters of debate.[3] The natural history of animals, especially insects, was not primarily a taxonomic endeavor. My concern throughout will be to discover what went into making and consuming natural history and how these activities fit into the lives of those who found the details of the natural world around them so engaging.[4] What did natural history mean to those who participated in it in one way or another? What drew them into such intricacies as the mechanics of metamorphosis or the mating behavior of mayflies or the sex of bees? How did they tie these phenomena to pressing theoretical issues like spontaneous generation or materialism? What techniques did they develop for seeing what had never been seen before?

In practice, boundaries between the sciences were generally fluid in the eighteenth century. Most fields, with the exception of medicine, required no particular training regimens, and men of science often moved among different subjects, from mathematics to astronomy to chemistry to botany. Many others, like Pierre Lyonet, pursued their scientific interests only when their other obligations allowed them the time to do so. This is not to say that there were no specialist traditions in these fields; certainly much of the knowledge produced in any of these areas was opaque to outsiders. But even when institutional divisions reified subject areas, as in the classes of the Paris Academy of Sciences, academicians frequently crossed these divisions between fields when choosing their research topics and staking their claims. In any case, the Academy of Sciences had no designated class of natural history, although natural historical observations and experiments found their way into virtually every meeting and every issue of its journal. Many of these reports were communicated by distant correspondents, making the institution a vital node in the informal network of naturalists and travelers worldwide.

Natural history was closely tied to the mission of the Academy of Sciences to test, authorize, and initiate useful knowledge of all kinds. It was also part of the larger culture of curiosity, amusement, and polite learning shared by urban and provincial elites. Partly because natural history encompassed so many areas and kinds of knowledge, it engaged readers, collectors, and observers without appealing to all of them for the same reasons. At the academy, natural history brought the institution into contact with its public, as systematic studies of particular creatures, substances, or phenomena took their places in the pages of the annual journal alongside eclectic and singular observations of curiosities or rarities, often sent in by distant correspondents. The semiannual public assemblies usually included a presentation on some natural historical topic, which would then be reported—often in considerable detail—in the press.

The central figure in this book is René Antoine Ferchault de Réaumur, preeminent naturalist of the generation before Buffon and a commanding presence on the scientific scene in Paris from the 1720s until his death in 1757. In exploring Réaumur's world—his social relationships, his intellectual production, and the natural world he investigated so vigorously—I develop a picture of natural history that revises and complicates its characterization as a science

Figure 1.1. R. A. F. de Réaumur, in formal dress, ca. 1735. Engraved by Philippe Simonneau from a painting (now lost) by A. S. Belle. Wellcome Library, London.

of straightforward description, devoted to amassing observations as the raw material for taxonomic systems. In his classic work on the life sciences in the eighteenth century, Jacques Roger presented natural history before the landmark work of Georges-Louis Leclerc de Buffon as predominantly descriptive, rather than theoretical or interpretive, with an emphasis on the structures and mechanisms of animal bodies. Roger divided naturalists in this period into "classifiers" and "observers"; in this scheme, Réaumur was the prototypical observer, mired in providentialist theology. Roger contrasted Réaumur's style of natural history to the philosophical, high-Enlightenment work of Buffon, whose sweeping gaze extended back to the beginning of time, across the ocean to the Americas, and down into the submicroscopic realm of forces and organic molecules.[5] This historiography unabashedly privileges theory and synthesis, reducing natural history before Buffon to plodding catalogs of structures, with no room for general explanations or scientific laws. Réaumur, whatever his merits as an observer, could not break free of "the mentality [*esprit*] of the age," in effect blinkered by his theological orthodoxy.[6] "For the observers," according to Roger, "the structure of things is interesting for its diversity and for its perfect 'fit' with the way of life and the milieu of each species. . . . And the observer delights in seeing each insect use to perfection the tool that Nature gave it to do what must be done to survive and reproduce. But neither the observers nor the classifiers are much interested in 'physics' [*la physique*], that is to say plant or animal physiology. . . . By its very definition, according to the terms of the time, 'physics' is excluded from natural history."[7] With no interest in physics, observers limited themselves to describing structures or mechanisms and admiring their design. This radical bifurcation between natural history and physics, based on the classical distinction going back to Aristotle between description and causal explanation, seems to miss the central character of natural historical investigation in our period. If we can avoid reading Réaumur through Buffon's eyes, we will see that natural philosophy and experimental physics both entered crucially into natural history.[8]

A close examination of the actual work naturalists were doing reveals the integral and sometimes surprising role played by experiment and measurement, through the direct application of techniques and instruments typical also of chemistry and physics. Marc Ratcliff's research on the extensive networks of microscopists across Europe and Britain in the eighteenth century effectively collapses the distinction between observation and experiment, and explores one of the experimental aspects of natural history. By zeroing in on the microscope, he shifts his perspective from philosophy to practice, without losing sight of the philosophical ramifications of what observers saw with their instruments.[9] In fact, naturalists often described their work as part of "*la physique*," meaning physical science broadly construed, or what in English was then called natural philosophy. This is a crucial point for understanding not just the methods used by naturalists but their conception of the whole enterprise. We even find Réaumur pointing out the common "spirit" shared by natural history observations and the techniques of the mathematician, to remind his readers that the naturalist does much more than describe and catalog whatever wanders across his field of view:

> The spirit of observation, the kind of spirit essential to naturalists, and commonly assigned to them, is equally necessary to progress in every other science. It is the spirit of observation that causes us to perceive what has escaped others, that allows us to grasp the relations among things that appear different, or that causes us to find the differences among those that seem similar. We resolve the most difficult problems of mathematics only once we can observe relations that do not reveal themselves except to a penetrating and extremely attentive mind. Observations make possible the resolution of problems in physics as in natural history—because natural history has its problems to solve; it even has a great many that have not yet been resolved.[10]

How a wasp constructs its nest, how dragonflies mate, or how a clam propels itself are examples of such problems, and the solutions must come from watching the animals at work, rather than speculating about hidden structures or even about final causes.

Réaumur started out his scientific life as a mathematician; the parallels he finds among problem-solving in mathematics, in physics, and in natural history underscore not only some unexpected shared ground linking these fields, but also the pragmatic way that naturalists framed their questions. If we follow Réaumur in thinking of natural history as problem-solving, we are on the way to a richer conception of what it meant to do this kind of science. Some of the problems were indeed mechanical, as Réaumur notes. He includes under this rubric the operations by which an insect builds its cocoon, lays its eggs, changes its skin, or traps its prey—none of which could provide a basis for structural classification. To see the answer to a problem, the naturalist has to know how to look, how to be attentive, and very often how to devise ways of looking that will uncover hidden maneuvers or mechanisms—like the respiratory apparatus of the insect described by Lyonet at the beginning of this chapter. By observing the habitual lifting of the tail to the water's surface in many different species, and then testing the consequences of preventing this behavior, the naturalist can not only describe but explain why the insects behave as they do. As in mathematics, finding the right point of entry into a question takes not just training and skill, but long experience.

The written histories of different species incorporated as much knowledge as possible—including anatomy, physiology, and behavior, as well as the useful consequences of this knowledge. Understanding the habits and physiology of a pest, for example, could inspire experiments on how to control it. The contemporary term "history," which I use here in preference to description, emphasizes the dynamic and fluid aspect of these accounts, which often move from anatomical details to narratives of mating behaviors or metamorphoses.[11] Writing these histories, and illustrating them, entailed sustained programs of experimentation and manipulation. Naturalists developed techniques of microscopy, dissection, feeding, confining, breeding, and simply watching and recording, all tailored to the various species that crossed their paths every day. When the naturalist wrote up a "history"—or problem solution, in Réaumur's language—he not only reported what he saw, but regaled his reader with tales of the procedures, tricks, and techniques that made the observation possible. These texts were histories of human investigations and histories of animal life at the same time, so that the hero of the narrative could be the naturalist or his subject, or both, depending on circumstances. The attentive reader of natural history books could learn as much about how to be a naturalist as about the habits and attributes of a particular insect or worm or spider.

INSECTS AND NATURAL THEOLOGY

The eighteenth century witnessed an efflorescence of providentialist natural theology drawing on many areas of natural knowledge. Book buyers indulged an apparently insatiable appetite for works of natural theology that used examples from natural history to reveal God's hand in the tiniest details of nature.[12] But while insect enthusiasts in the eighteenth century were hardly averse to final causes or to seeing the handiwork of God in nature, most of them were not primarily motivated by theological considerations. Naturalists took for granted the existence of design and fitness in insects—a category in which they included everything from coral to worms to mites to aphids to butterflies and bees.

One revealing example of the complex relations between physicotheology and natural history can be found in the book where Lyonet's comments on water bugs appeared. This was the French translation of Friedrich Christian Lesser's *Insectotheologie*, originally published in German in 1738.[13] Lesser, an erudite Lutheran pastor, amassed bits of evidence and arguments from sources ranging from the church fathers and the Bible to natural history works of every era, in order to demonstrate divine providence in the insect world. Before printing a French edition of Lesser's work, the Dutch publisher asked Lyonet to look it over. Lyonet came from a Huguenot family and worked in The Hague as a lawyer, as well as a cryptographer and translator for the government. He was also a gifted artist, a collector of shells and fine art, and a fanatical observer of insects.[14] Though he had not yet published anything of his own when asked to review the Lesser translation, Lyonet was well known locally for his ongoing study of the insects of the region. Writing to Réaumur, he described Lesser's book, sardonically, as "a work composed in the German style, where pedantic work and reading play a greater part than genius and direct experience."[15] As he made his way through the text, Lyonet could not restrain himself from correcting numerous factual errors and misapprehensions and adding extensive footnotes, sometimes so voluminous as to displace the original text entirely from the page.

In his annotations and explications, typographically distinct at the bottom of the page, Lyonet adopted the persona of a working naturalist, playing off the foil of the learned pastor. Lesser had culled his evidence from various authorities without distinguishing between them or evaluating them; Lyonet drew on contemporary natural history literature but frequently supplied his own very detailed observations to challenge generalizations and misstatements. His religious convictions were entirely in sympathy with those of the German author, but he objected to the sloppy use of natural knowledge to argue for God's providence. Natural theology ought to be held to the same rigorous standards as natural history. In his notes, Lyonet sometimes simply corrected mistaken descriptions; sometimes he elaborated on a point made in passing in the text, explaining what he had observed in relevant cases; sometimes he undermined one of Lesser's generalizations with detailed counterexamples; and sometimes, strikingly, he introduced his own radically new discoveries. In effect, Lyonet hijacked the original text and claimed it for his own style of natural history, a style very much informed by his reading of Réaumur. The first-person narratives of what he had seen with his own eyes or understood as a result of his own experiments contrasted with Lesser's compilation from written sources, many of dubious credibility. "In matters of natural history," Lyonet admonished his reader, "it is dangerous to admit the marvelous on simple hearsay; but neither must we reject everything marvelous, because it seems unlikely to us. One must examine nature, and pay attention to the proofs [of such marvels]."[16] If an inexpert observer reports something without saying how he knows it, or how he was able to see it, or how he performed the experiment, the reader is

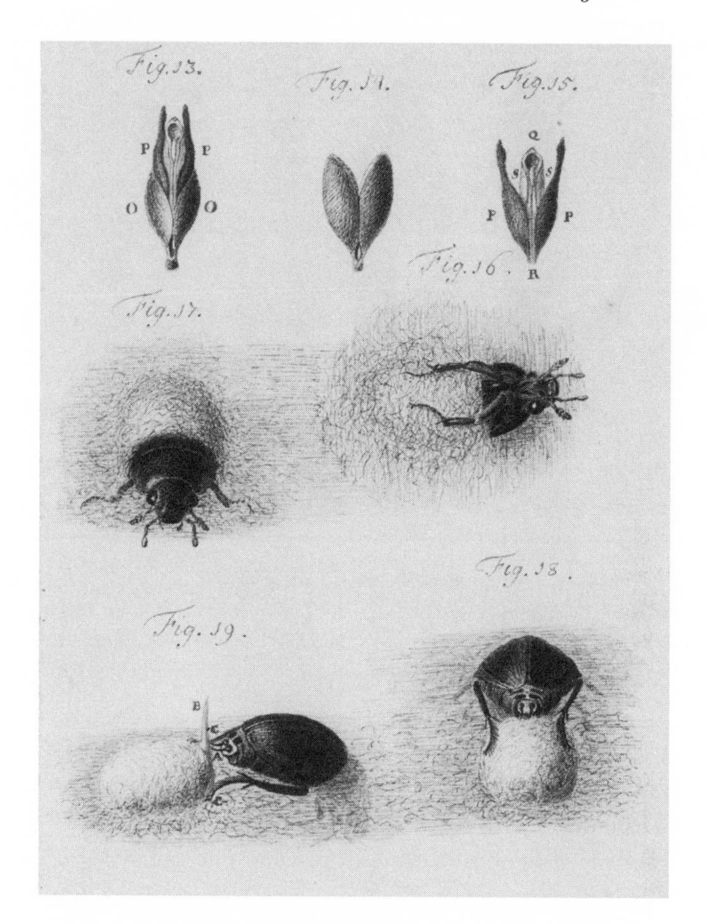

Figure 1.2. Pierre Lyonet, drawings of aquatic beetle, observed in captivity: male reproductive organs (figs. 13–15); female beetle constructing egg case (figs. 16–19). Enclosed in Lyonet to Réaumur, 17 May 1743. Ms. Fr. 99, Houghton Library, Harvard University.

✲

justified in withholding acceptance of the result. If, however, a "faithful and enlightened" author reports something extraordinary like the regeneration of a crayfish claw, if he explains that he cut off the claws of numerous crayfish, kept them under observation in jars, and charted the growth of the new appendages, "I should believe it, however marvelous it appeared to me."[17] (The example here was an implicit reference to Réaumur's study of regeneration in marine animals.)[18] After completing his extensive notes, Lyonet convinced the publisher to include in the French edition two elaborate plates packed with

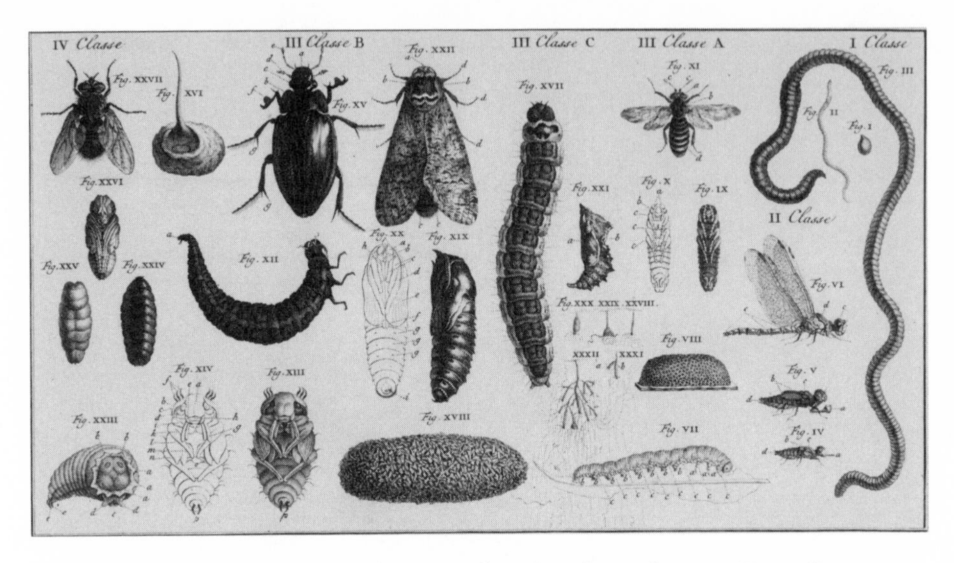

Figure 1.3. Lyonet's revision of Swammerdam's four classes of insects. "Drawn from life" by Pierre Lyonet; engraved by J. van der Schley for F. C. Lesser, *Théologie des insectes* (1742). Corrected proof copy of the plate enclosed in Lyonet to Réaumur, 22 February 1743. Ms. Fr. 99, Houghton Library, Harvard University.

✲

meticulous illustrations drawn by himself, which only underscored the immediacy and legitimacy of his eyewitness accounts of insect behavior.

By piggybacking on Lesser's earnestly pedantic physicotheology, Lyonet undertook a kind of proselytizing of his own, for an observational natural history that took nothing on faith, that looked beneath the surface, and that could ultimately be made into more convincing and more legitimate evidence for divine wisdom. Many of the good pastor's mistakes had nothing to do with the validity of his providential argument—they were mistakes of natural history, not of theology, and his commentator would not let them pass unnoticed. When Lesser mentioned that caterpillars lack ears, for example, Lyonet pointed out that caterpillar ears would not necessarily look like the ears of other animals, and it might not be so obvious how to recognize them if they did exist. In any case, he went on, we know next to nothing about the caterpillar's ability to detect sounds. When Lesser asserted categorically that all living creatures breathe, Lyonet challenged the universality of respiration with evidence from his own experiments. In one of these, he took a large chrysalis, covered its stigmata with soapy water, and watched through a magnifying lens to see if soap bubbles would form as the chrysalis breathed. "No matter how much attention

I devoted, I never saw anything of the sort."[19] Then he tried carefully placing a bubble over each pore and watching to see if the bubbles expanded; they did not. In another case, when Lesser noted that insects lack intelligence, Lyonet insisted to the contrary that his own close observations had shown how insects can vary their actions according to the situation, and even, "when placed in troublesome circumstances, where they would never naturally find themselves in the ordinary course of things, we see some that always make the best choice, and that know how to recover from mishaps and pull themselves out of very difficult situations."[20] The naturalist, placing the insects in these situations to see what would happen, is irresistably "tempted" to assign some degree of reason to these creatures.

In practice, in the heyday of physicotheology, natural history drew on much more than a reflexive providentialism. The reports of observers and experimenters, even those committed to providential theology, rarely put design arguments at center stage, and often were remarkably understated in this regard. The narratives and illustrations embedded in the effusive details that filled natural histories of insects yield many kinds of clues about a lively scientific practice that cannot be reduced to, or dismissed as, natural theology. This is not to deny that insect-watchers sometimes rhapsodized about the divine workmanship on display in the insect world. But this kind of physicotheology is often only the bare scaffolding surrounding the substantial edifice built from long hours of observation and experiment. As in the footnotes Lyonet offered as correctives to some of Lesser's uninformed claims about evidence for God's providence, that sustained dedication outweighed theological arguments, without necessarily challenging them.

The most accomplished and committed naturalists were driven by passionate fascination with their small subjects, often enriched with the conviction that knowledge of these elusive creatures might have practical as well as intellectual benefits. And beyond possible practical benefits—medical remedies, say, or methods for pest control, or new sources of textile dyes—they maintained a general but profound commitment to "the progress of knowledge," a sense that every new history of another species, whether mundane or exotic, contributed to the ever-expanding stock of knowledge about nature. The demonstration of God's wisdom was more of a by-product than a motivating force, as naturalists used whatever methods they could devise to allow them to understand how spiders spin their silk or lizards grow new tails or dragonflies fertilize their eggs.[21] "In the painstaking observations of the naturalists," Lorraine Daston points out, "the argument from design became less an argument from evidence than an experience of self-evidence."[22]

NATURAL HISTORY IN PARIS

In Paris, natural history clustered around two institutional nodes, the Academy of Sciences and the royal botanical garden, the Jardin du roi. Most of the botanists in the academy had some affiliation at the garden, as did several of the chemists. There was considerable cooperation between academy and garden, usually mediated by personal friendships and patronage ties. In spite of the overlap in personnel, there was also a certain amount of tension and competition between the two institutions, and this intensified after the appointment of Buffon as director of the garden in 1739.[23] The francophone community of people observing insects and trading specimens, results, and drawings extended far beyond the borders of the Paris institutions, however. The botanical garden maintained a network of travelers and correspondents to supply its greenhouses and seed collections and other cabinets, and the academy also cultivated connections with naturalists reporting from distant locations, appointing particularly useful contacts as official corresponding members.[24] In addition to the experiments and observations produced locally, the academy became a clearinghouse for new observations of all kinds. Reports and specimens from distant places were presented at meetings and formally evaluated by academicians; some of them were then published in the annual journal. The publicity the institution could offer through its widely disseminated publications gave provincial naturalists or collectors living in colonial or commercial outposts an incentive to communicate novel, curious, or potentially useful observations. At the same time, many people pursued natural history far from the capital and communicated with each other locally, through networks that could bypass the capital entirely. Outside of France there were many other nodes of intensive natural historical activity, again with their own networks. Geneva was one such center, the Netherlands had several others, including The Hague, where Lyonet worked. Farther afield, in places like Madagascar and Cayenne, observers and collectors had their own operations. They make their appearances in our story when they communicate with Paris, but they had their local circles of exchange and collaboration as well, activities that usually left only sporadic traces in the archives.

The Parisian naturalist Réaumur, a prolific academician, proprietor of an ever-expanding collection, and the author of a multivolume work on insects, actively recruited collaborators and contributors to his various projects. In the service of his collections and his research programs, he maintained a voluminous correspondence, sending books and instruments across Europe and around the globe in exchange for specimens and letters. This correspondence gives considerable insight into the practice of many kinds of people, their various interests, motivations, and methods, as well as the questions they were asking. Some of them were looking for support for their own publishing ventures; some volunteered to contribute to Réaumur's collections and to the academy's journal; others were recruited for their special skills or for their access to distant locales. The observations and experiments presented directly to the academy represent only the tip of the iceberg of natural historical investigation in the period. I cannot pretend to survey all of the venues and all of the people engaged in these practices; instead I will draw on an eclectic sample of letters, published papers, and books to discover how this diffuse community operated, and to retrieve the meanings and values its members attached to their work.

KNITTING WITH SPIDER SILK

As an introduction to this mode of engaging with the natural world, I turn to an example from the first decade of the eighteenth century, when papers on insects and other small creatures started to crop up with increasing frequency in the minutes and publications of the academy.[25] These investigations often began as chance observations of odd phenomena, or as evaluations of novelties sent in from other places. The strength and use of the silk of spiders was brought dramatically to the attention of the Parisian academicians by a communication from their sister institution in Montpellier, the Royal Society of Sciences.[26] In December 1709, François-Xavier Bon de Saint-Hilaire read a paper on "the usefulness of spider silk" to the public session of the Montpellier academy, where he was an honorary member, and then sent it on to Paris. Bon belonged to the lower nobility and served as a magistrate in Montpellier.[27] His paper reported on experiments on the fiber from the egg cases (*coques*) of short-legged spiders. He had found that the fibers could be made into serviceable thread "no less beautiful than ordinary silk," which could even be dyed. He passed around a pair of stockings and a pair of mittens knitted from spider silk, and he ended with a chemical analysis, displaying the distillation products and comparing their properties to those of silkworm cocoons. Bon also reflected on the benefits of natural history for someone like himself, who needed some respite from the press of his daily work. He did not need to remind his local colleagues of his considerable experience with natural history investigations in the years before he took up his hereditary position in Montpellier's judicial administration. As a young man, Bon had worked with the famous Italian naturalist and adventurer, Count Luigi Marsigli, on his study of coral in the coastal waters near Marseilles. This had included chemical analyses similar to the distillations of spider silk.[28]

Bon's paper is much more than a description of spiders and their egg cases. The magistrate-naturalist had evidently devoted himself to watching spiders spin, examining their anatomy and physiology to discern the mechanics of silk production, as well as collecting, processing, and experimenting with the silk itself. He had lived with spiders over many months, pursuing a sustained investigation, with the help of anonymous assistants of various kinds. Curiosity, amusement, and utility all motivated his work. In order to determine whether spiders could be cultivated viably as a source of silk, Bon needed to know more about their life cycle and habits. "I ordered all the large short-legged spiders found in the months of August and September to be brought to me. I enclosed them in paper cones and in pots; I covered these pots with paper pierced with several jabs of a pin, as I did with the cones, so that they would have air. I had flies given to them [to eat], and I found some time later that most of them had made their egg cases." He obtained large quantities of additional egg cases simply by offering to buy them by the pound, just as one would acquire ordinary silk from silkworm cultivators. Thirteen ounces of the spider shells yielded four ounces of clean silk, more than enough to make a pair of stockings for "the largest man."[29]

Bon's spider paper has all the features of a typical academic memoir on a natural history subject. His methods—including direct observation, laboratory analysis, quantitative comparisons, and even the production of artifacts—put him squarely in the mainstream of natural historical practice. He systematically observed a variety of spiders, came up with a rough categorization (dividing them into long- and short-legged species), discovered their eating and mating habits, and kept them in captivity while they produced their egg cases. He performed calculations as well, comparing spider silk to the more familiar fiber made from silkworm cocoons. Although he did not conclude definitively that spider thread could be economically profitable, his comparison to silkworms was favorable on several counts, and he imagined that spider cultivation could yield practical benefits, represented materially by the stockings and gloves he showed to his confrères.

Bon was sufficiently pleased with his results to communicate them immediately to the Royal Society in London, where an English translation of his paper appeared in the *Philosophical Transactions* for 1710, as well as to Paris.[30] Exploiting his family connections, he wrote directly to Father Jean-Paul Bignon, the state minister responsible for the royal academies. Bignon asked Bernard de Fontenelle, perpetual secretary of the Paris Academy, to read a summary of Bon's letter to the assembled academicians only a week after their counterparts in Montpellier had admired the spider-silk stockings. Fontenelle drew attention to the elaborate patronage chain traversed by the knitted curiosities. "M. Bon . . . found in the spiders a strong and lustrous silk, of which he made a pair of stockings that he gave to the Duke of Noailles, who is supposed to present them to the Duke of Burgundy, and some gloves made specially for Father Bignon. He says also that he extracted from this silk a spirit and a volatile alkaline salt, stronger and more active than that of ordinary silk."[31] In response to Bon's letter, the Paris academicians designated two of their number to conduct their own sustained investigation of spider silk.

One of these was the ambitious young mathematician Réaumur, only recently appointed as *élève géomètre*, not yet established as a naturalist, and not yet a powerful figure in the academy.[32] He took the spider assignment seriously, and reported on his own experiments to the public session the following November.[33] The stockings and gloves sent from Montpellier were "incontestable proof" of the possibility of knitting with spider silk—but could this material be commercially viable? Réaumur recounted his attempts to raise spiders, and his observations of their habits and physiology. He approached the spiders and their egg cases as a problem of economic natural history, setting up his trials to afford comparison to the silkworm. He would first attempt to raise large numbers of spiders, and then investigate whether their silk could be made more cheaply than conventional silk—"or, if it is more expensive, to see if that drawback would be compensated by some other advantage."[34] Enclosing large numbers of spiders in glass-lidded boxes, he watched them feast on segments of earthworms. They also liked the soft material from the inside of the quills of young pigeon feathers. He tried feeding this to his spiders because it resembled "the soft and tender matter that they like to suck out of insects." He described this readily available spider food rather casually: "Anyone who has taken the trouble to squeeze [the quill of a young pigeon feather], or to dissect it, will have found it to be filled with a tender substance crisscrossed with a great number of vessels."[35] He harvested this delicacy and found that the spiders, especially the young ones, ate it happily.

Although he politely acknowledged Bon's contribution, Réaumur implied that his own investigations were more systematic and therefore more conclusive. He watched large numbers of spiders hatch and work together on a single web, but this did not last long. "I put two or three hundred in some boxes, in others a hundred, or fifty, or even fewer; these boxes were about the size of a playing card . . . a large enough space for such little animals." But as they grew bigger, they did not live together so peaceably, and soon they were devouring each other; if they were instead isolated from each other, the problem of feeding them individually became overwhelming. In addition, Réaumur examined the process of spinning in different species, noting that the silk of spider webs differed radically in thickness from the fibers destined for egg cases. To test the strength of the fibers, he attached weights to them and found that the thicker thread of the egg cases could support about eighteen times as much weight as the web threads. Silkworms, however, spun even stronger silk, by a factor of five.[36]

Finally he compared the output of the spiders and the silkworms: "I very carefully weighed many silkworm cocoons and I found that the strongest . . . weighed four grains, and the weakest weighed more than three; . . . [so] it takes at least 2304 worms to get a pound of silk. . . . With the same care, I weighed a great number of spider shells, and I found that about four of the biggest were needed to equal the weight of a silkworm cocoon."[37] In the end, he determined that it would take "about" 55,296 spiders to produce a pound of silk, and they would have to be kept in little compartments and fed individually for several months to keep them from killing each other. The silk of spiders would thus become incomparably more expensive than that of silkworms, largely because of the cost of the labor required to raise them. Regretting that such an "ingenious discovery" could not be practical, the paper nevertheless ended on an optimistic note, suggesting that perhaps American spiders, of which several large species were known, might be more viable as silk-producers. "Whatever the case may be, we must experiment; that is the only way to discover curious and useful things."[38]

Bon's reaction to these conclusions survives in Réaumur's papers, apparently copied from a letter he wrote to another academician: "I do not find his calculation correct," Bon complained. "He forgot that each spider produces at least eight hundred to nine hundred new spiders, while each silkworm produces at most one hundred, of which only about thirty survive." He also disagreed about the inevitability of the spiders killing each other: "The best reason one can give to prove that spiders do not kill each other when they are at liberty, as for instance in a room, is that one finds a great many of them, and that the egg cases are very close to each other."[39] Though he was not convinced that spider silk could ever compete on the market, Bon did not think Réaumur's conclusions as compelling as his numbers seemed to suggest, and he did not like the way his own research had been co-opted by his Parisian colleague. Bon kept his work in the public eye by reprinting it several times. Both Bon's and Réaumur's academic papers were extracted in the *Journal de Trevoux*, and provided the Jesuit missionary Father Parennin with subject matter for his conversations with the Chinese emperor Kangxi.[40]

The study of spider silk nicely displayed the intertwined and essential elements of natural historical knowledge: economic (cost of labor and materials), physical (quantitative measures of fiber weight and strength), curious (the stockings), physiological (spinning), anatomical (structure of the silk glands), and behavioral (eating habits, including the tendency of larger spiders to eat smaller ones in captivity). Naturalists often incorporated methods and techniques from the physical sciences—especially chemistry, but also physics, or what later became known as experimental physics. Though they disagreed on particulars, Bon agreed with Réaumur in practicing natural history as physical science: "All philosophers, and especially the moderns," Bon asserted, "have regarded this science [natural history] as the foundation of physics. If they dedicate themselves to searching with exactitude for certain facts, it is only to arrive subsequently at the true knowledge of causes."[41] Here Bon showed his familiarity not only with chemistry but with the tradition of Baconian empiricism that valued natural history as the foundation on which other sciences built.[42] After collaborating with Marsigli on coral, Bon put aside natural history to take up his position as magistrate in the Montpellier judicial courts, keeping the sciences for his leisure time. Réaumur, on the other hand, renounced life as a provincial seigneur, refusing his hereditary role as pater familias and allowing his younger brother to live on the estate and manage the property. He made a career in the capital, not as a landowner, but as an unmarried academician.[43] Though he and Bon chose different paths in life, they came from similar social backgrounds, and shared many of their interests and pursuits—not just spiders, but collections (minerals, fossils, shells, animals, insects) and a passion for measurement.[44]

Réaumur's major work on insects, the six volumes of *Mémoires pour servir à l'histoire des insectes* (1734–1742), did not begin to see print for almost twenty-five years after he completed his experiments on spiders and their silk. In the interim, he alternated work on insects and marine animals with investigations of many areas of applied science, including new methods for forging steel, producing porcelain, and making paper. For a few years during the regency, he supervised the "Enquête du régent," a survey of local industries and natural resources in all regions of France, and he was also responsible for the academy's long-standing project to collect and publish descriptions of artisanal techniques.[45] He designed a thermometer that became a standard scientific and meteorological instrument all over the world for generations.

He experimented on the regeneration of severed appendages in crayfish and starfish; he studied fossil shells, coral, and the properties of metals; he observed wasps and honeybees and ants. By the 1730s Réaumur was established as the reigning expert on insects and moved into other areas as well: reproduction in frogs, digestion in birds, taxidermy techniques, and eventually poultry breeding and the artificial incubation of chicken eggs.[46] His energetic work in the first half of the century arguably set the standard for observational natural history for many years to come. He enlisted other naturalists and various assistants and amateurs in experimental programs prompted by contemporary controversies about generation. His work on insects and chickens sparked enormous interest in observation, collecting, and experimentation among his readers, who included provincial intendants, colonial and provincial physicians, military officers, noblewomen, clerics, nuns, and members of the landed gentry. Réaumur's books also inspired the investigations of young men who subsequently made natural history their own work: Charles Bonnet, Abraham Trembley, Pierre Lyonet, Charles de Geer, Jean-Etienne Guettard, Henri-Louis Duhamel du Monceau, Gilles Bazin, and Mathurin Jacques Brisson.

In justifying his focus on the insect world, Réaumur compared natural history to celestial physics, "the object of sublime speculations and worthy of the greatest minds."[47] Some might think that a naturalist who spent his life studying insects was wasting his time on trivialities. At first glance, very small objects may seem less worthy of awe and attention than the stars, but their movements, and their lives, are no easier to understand. "It may be more difficult to explain the causes of the movement of fluids in insects, preparations and filtrations of the material that becomes silk in the organs of some of them, the action of their stomach, the play of their admirable lungs, the growth of these insects, the shedding of their skins and their transformations; it may be more difficult to find the cause of the motion of the least muscle than to find the cause of the motions of the celestial bodies."[48] Far from trivial, as a challenge to observation and explanation the insect world was thus prime material for the most adept *physicien*.

Seeing into the unexpected intricacies of the very small, and then figuring out how to display these details as spectacles in their own right, inspired all sorts of people to turn an inquiring gaze on insects. In many cases, these observations became part of a larger natural historical practice: making collections, experimenting with preservation techniques, microscopy, raising and breeding birds, and the like. My goal in this book is to show how observing, collecting, and experimenting fit into the daily lives of a diverse array of people in the mid-eighteenth century. In order to do this, I reconstruct practices and decipher motivations in the loose networks of friends, associates, collaborators, readers, artists, and travelers who made natural history possible. I approach this material in the spirit of recent scholarship in the history of science that has emphasized practice and place.[49] Though Réaumur is my main character, by virtue of his prolific output and his position of power in Paris, over the course of the chapters to follow, I also follow his correspondents to fashion a portrait of a variegated and dispersed community of people who valued this kind of natural history and devoted substantial time and effort to doing similar work themselves. Throughout this book, I quote frequently from my primary sources for examples of the maneuvers of insects, the behavior of chickens, or the interventions of their observers. My illustrative examples, with their fine-grained detail, stand in for a stunning profusion of similar material, in an effort to bring natural history alive in its particulars, and to transmit something of the flavor of these investigations. Eighteenth-century readers would have encountered these observations embedded in much more densely textured accounts, overflowing with more circumstantial details, referring back and forth to other texts and debates, and extending over many more pages. I have tried to recuperate this world without reproducing it.

2

"Catching Nature in the Act"

IN THEIR QUEST TO SEE every detail and every phase of insect life, many observers lived with their subjects, day in and day out, "raising caterpillars . . . in the same way that we raise domestic animals."[1] The insects had to be fed, kept at the appropriate temperature, and in some cases supplied with the raw materials for their nests. A butterfly's eggs could be kept indoors and watched for signs of life, as in this typical report: "I had a butterfly . . . that laid its eggs where I live [*chez moi*]. They were successfully fertilized and hatched around the 25th of July. As their first meal these little animals ate the shells of their eggs. They made webs around the piles of eggs to be sheltered there, in order to eat at their leisure. They only went to the rotten wood I left for them after having absolutely dispatched their shells."[2] The accumulation of curious details, such as the web constructed to screen the tiny caterpillars as they made their meal, formed the core of natural history practice. This chapter explores the daily routines and improvised strategies of those who spent time, often considerable stretches of time, watching and recording the lives of insects. These were not, for the most part, exotic species discovered in distant lands and brought back for European collections, but rather the common inhabitants of backyards and forests, gardens and barns, roadside hedges and coastal tidepools: homely caterpillars, spittlebugs, aphids, spiders, mites, weevils, and inchworms. Naturalists shared a commitment to watching live insects, tracking their various maneuvers, and following them through the stages of their life cycles. They counted as discoveries any unknown detail of structure, physiology, or behavior, seeing what had never been seen before, even in the most familiar insects. They were also, in general, conversant with the existing literature on insects, and often referred to their books as well as to their magnifying lenses and their boxes of bugs and butterflies.

The naturalist deciphering the identities and actions of caterpillars and butterflies faced an embarrassment of riches, repeatedly coming up against the problem of how to see ephemeral events among creatures whose size and habitat normally kept them out of view. Important as it was to observe insects in their native surroundings, to know what they ate and how they lived, systematic investigations meant bringing them indoors and keeping them alive, while keeping them visible. "When we leave caterpillars at liberty in the countryside, when we observe them only on the plants they like, it is only by good fortune that we can succeed in seeing them make their cocoons, or often even in finding their cocoons, since most of them leave the plants they have always lived on to go and spin in other places. To follow them in their work, there is nothing better than to raise them in enclosed spaces, and especially in glass jars, which allow us to see them at every instant."[3] Even enclosed in a jar, with the proper food, a caterpillar might not spin its cocoon or a butterfly lay its eggs in the absence of conducive conditions. A caterpillar found on a nettle attracted Réaumur's attention with the noise it made chewing on the paper cover of the jar where it was preparing its cocoon. It was detaching bits of paper and carrying them to the cocoon "which it wanted to cover with carefully arranged pieces of paper, one next to the other, apparently to substitute for the material that its species would find in the wild, and which I did not know to provide for it." The observer immediately started lowering strips of paper, impaled on the end of a pin, into the container so the support for the cocoon would not be shredded by the caterpillar.[4] In another case, when a particular kind of butterfly would only lay its eggs in the dark of night, he put them in square glass boxes and covered them completely, so no light could get in. After a few hours, taking off the cover, he was able to surprise them in the process of emitting the packets of eggs.[5]

"Following insects in their work" makes a nice description of the naturalist's own work; following the naturalists as they work is what I propose to do here. What went into this kind of work? What questions and assumptions inspired these observers to their feats of minute inspection and experimentation? As we peer into the lives of insect-watchers, we will notice that sociability played a crucial role in refining techniques and exchanging and understanding specimens. Many observations were made in company, whether with assistants, family members, or friends. During a walk in the Vincennes forest, one of Réaumur's "insect-hunters" presented him with a twig and a caterpillar just starting to make its cocoon. "I watched this caterpillar work to enlarge and raise these two strips for more than an hour and a half. I showed it, occupied with its work, to several people who were walking with me. We were careful not to shake the little branch where [the caterpillar] was, . . . and it continued with its task as we watched."[6] The outing, in the forest not far from Réaumur's country house just outside Paris, drew these friends into the observation of the caterpillar's construction, which then led the naturalist to reflect on the question of intelligence in insects. Though we cannot listen in on the conversation, the story of the walk in the woods gives us a glimpse of the kind of sociable excursion where many bits and pieces of natural historical knowledge originated. Even solitary investigations, if successful, could fuel conversations and demonstrations to guests in various domestic settings. Discoveries made in gardens and fields, as well as studies, cabinets, and laboratories, might then be taken to the academy or exchanged with distant correspondents. Many of the observers who played integral roles in making natural history never entered the meeting room of the academy, even if their observations did. In the vignettes of scientific life that follow, I pay particular attention to how knowledge was made locally and circulated in correspondence, before it moved into the more formal space of the academy's meeting room and perhaps on into the more public world of print.

The tricks of the trade developed by naturalists shifted and evolved constantly to suit the species under observation. Watching for the emergence of a nocturnal butterfly from its chrysalis, for instance, one might try arranging as many of them as possible within the reach of a lantern's light. In years when they were plentiful, Réaumur found he could collect a large number of such chrysalises and pin them to the tapestry in his study, where he could keep them in view as he waited for them to hatch. "Each one was fixed there with a pin passed through the leaf it was attached to, or with a pin passed through only the attaching threads. . . . With such supplies of chrysalises, which only remain in that form for fourteen or fifteen days, I often saw butterflies hatch, without being obliged to lose time waiting for the moment of the observation."[7] The array of chrysalises and cocoons hanging on the wall accommodated the working habits of the naturalist, who could sit at his desk doing something else until the motion of a butterfly or moth drew his attention. In this way, the study was transformed into a kind of observatory where the diminutive creatures could not escape the naturalist's gaze—as long as his attention did not wander too far.

This is only one among many, many strategies for bringing insects into the more or less controlled space of the study to make them accessible to observation, manipulation, and experimentation. The naturalist's enterprise, fluctuating in intensity and character throughout the year, and adapting to location and other contingencies, often put dozens or hundreds of specimens under scrutiny to optimize observing opportunities, and especially to exchange the realm of happenstance for something more conducive to witnessing an ephemeral event. So, Réaumur explained, in order to identify the butterfly form of a particular caterpillar, only one or two specimens would be needed, since the timing of the metamorphosis would not be crucial. "But if one wants to catch these caterpillars in the act of delicate operations that they perform only once in a lifetime, lasting only a few instants, it is chance if the moment of these observations does not escape the observer who has raised only one caterpillar of that species. If he has raised hundreds of them, he has multiplied by hundreds of times the occasions of observing these precious moments."[8]

Not all insect-watchers collected and raised their subjects on such a grand scale, but neither was Réaumur alone in adapting his observational habits to the lives of insects and adopting many species of insects into his household. In his publications, he referred frequently to the contributions of his correspondents as well as various associates and assistants observing in his company. Sometimes these references simply acknowledge the source of specimens or observations, but often the narratives include details about the circumstances of discoveries and the people engaged in accumulating not only specimens but insights and techniques. The eclectic style of Réaumur's chapters, which describe experimental programs as well as the insects themselves, show the activity of observation as part of daily routines. These texts can then be read alongside the letters and notes to reconstruct motivations and practices more fully.

My account of these activities is necessarily somewhat impressionistic. In most cases, I have only one side of a correspondence and many other letters are missing. None of the insect specimens or preparations survive. Occasionally sketches or drawings are still kept with the letters, but even when this is the case, they rarely represent apparatus and never show observations actually being performed. I am dealing with only a small, unscientific sample of the much larger number of people who actually watched, collected, pondered, prodded, dissected, and preserved insects. Many such observations were of course redundant or inconsequential. Rather than tracing the genealogy of particular discoveries, I focus on the many facets of the activity of insect-watching, as it played into the study of natural history, and as it fit into the daily lives of observers. In pursuit of previously unseen traits and phenomena, naturalists used a special kind of ingenuity to capture, feed, observe, dissect, and preserve their insects. Quite a number of Réaumur's suppliers and collaborators were in effect trained by him, if not in specific techniques then in honing the sensibility that allowed them to notice insects and contrive to follow their movements and their life cycles. They shared a fascination with what Réaumur called the "beautiful mechanism" of whatever creatures they were examining. They exchanged techniques for seeing into these hitherto uncharted corners of creation, regaling each other with tales of insect habits and anatomy, and egging each other on to new observational feats.

Réaumur cultivated connections that supplied him with a steady stream, sometimes an avalanche, of observations, ideas, and specimens. His modus operandi—in his own daily practice as well as in the habits he encouraged or inculcated in others—was to gather more or less indiscriminately, in order to have as much raw material as possible. Some things might be new; some things might supply a missing link in the understanding of a particular species; some might corroborate previous suspicions or allow repeated observations of elusive phenomena. Although many of his correspondents regarded him as an established master of natural historical observation, he did not control his informants any more than he controlled the natural world they were investigating. He made use of their observations and their specimens for his collection and publication projects; they made use of his expertise and his goodwill in various ways, depending on their needs and circumstances. The dynamics of exchange so crucial to this type of knowledge-making involved friendship and patronage relations in a complex dance of circulating objects, people, preserved specimens, and living creatures.

Writing letters was an integral part of the observational life of naturalists. Their letters condensed a great deal of activity, with descriptions of anatomy and behavior, of techniques and apparatus, and of local people and the contingencies that made particular observations possible. Though we might think of watching insects as a more or less stationary activity, in fact observers, their letters, and specimens were frequently in motion, through shorter or longer distances depending on circumstances. Letters overflowed with discussion of how to get things (boxes of insects, supplies, instruments, books, journals, drawings, as well as the letters themselves) from one place to another reliably and at feasible cost. Reading these letters, we get at least an inkling of the overlapping and intertwining movements of these things, the miscellaneous components of which natural historical knowledge was made.[9]

In the early 1730s Réaumur was solidifying his position as an expert on insects, contributing papers on the subject to the Paris Academy of Sciences, and continuing to amass material for his books, printed by the royal printing house over an eight-year period starting in 1734. I start this ramble through his correspondence with a circle of friends living in the region of Poitou, near Réaumur's estates, who got interested in caterpillars and other insects in

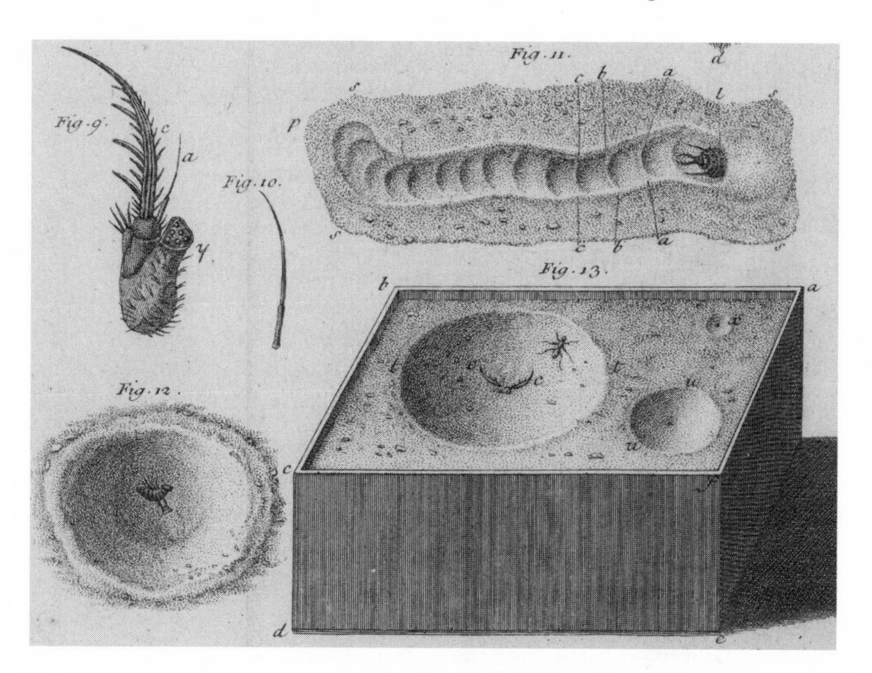

Figure 2.1. Catching ant lions in the act: constructing trap in sand (fig. 11); lying in wait for prey (fig. 13); grabbing ant (fig. 12). Engraved by J.-B. Haussard from drawing by Hélène Dumoustier. Réaumur, *Mémoires pour servir à l'histoire des insectes*, vol. 6, plate 32 (detail).

✲

this period. There was nothing spectacular about the subjects of these observations, brought from the hedges and fields of the surrounding countryside into studies or kitchens, where they were coaxed into spinning cocoons or laying eggs. Though there were many questions and uncertainties about these creatures, they were not as hotly contested or as publicly engaging as the questions debated a bit later, about asexual reproduction or regeneration in worms and hydra or the origins of infusoria. The ubiquity of caterpillars, their very ordinariness, made them accessible; the variety of species meant that there were endless new discoveries to be made. To those collecting, raising, observing, dissecting, and drawing these little creatures, such discoveries and encounters with nature were anything but trivial. Metamorphosis remained an enticing mystery, tying back to the problem of generation, always hovering in the background of these observations. The change of form was not the only thing that drew observers to insects, however. The seemingly endless variety of their behaviors and structures, the hidden mechanisms of their body parts, and the many cases where something like intelligence seemed to guide them all contributed to insects' appeal as objects of sustained attention.

EXPERIMENTAL SETTINGS

In Paris, and in his nearby country house at Charenton, Réaumur kept his research activities going with various kinds of assistance. He had a series of helpers in the "laboratory" in his Paris home; some of these worked on physics experiments as well as natural history observations (see chapter 3).[10] The continual round of experimentation, collection, observation, and illustration was very much a domestic enterprise, with a shifting cast of characters and frequent interaction with the more formal setting of the Academy of Sciences and the less formal setting of the surrounding countryside. Every September, when the academy took its two-month vacation, Réaumur traveled with a small entourage to the seigneurial village (also Réaumur) where the estates yielded insects as well as agricultural produce.[11] These annual sojourns fostered interest in natural history among the neighbors, and the estate became a node in a network of observers throughout the region, connected by personal acquaintance and family ties. This provincial circle had a special status, since it was made up of people who knew each other and who traded observations in person as well as at a distance, at least during some seasons of the year. The roster of Réaumur's correspondents included several of these friends, who continued to supply him with specimens from his native countryside even after he had returned to the capital.

One of these correspondents was a Dr. Baron, whose surviving letters span about fifteen years. Little is known about his life, apart from clues in the letters themselves and in brief references to his observations in print. Baron lived at Réaumur's Paris home for a time, probably while completing his medical education. His responsibilities included taking care of the "insect menagerie" before he left the capital to establish a medical practice in Luçon, not far from Réaumur's estate.[12] His experience as part of the household gave him excellent credentials as a provincial observer, and the two men continued to meet during the annual vacation. Baron took obvious pleasure in his collecting and observing activities as a respite from the grind of professional obligations. For a few years, he kept shipments of local specimens flowing toward Paris, participating vicariously in the learned world there. Eventually he married and gradually adapted to family life in the provinces, tending his garden and keeping track of the insects.

Baron found the transition from life in Paris to his provincial practice difficult, bemoaning his lack of time for natural history, an occupation he preferred to medicine. He had plenty of patients, but he complained that they were inordinately demanding and often did not pay for his services. In the early years, he yearned to return to Paris. He wanted to remain useful to Réaumur, sending boxes of specimens and observations of insects (especially garden pests), and he maintained ties to the circle of local savants who went on rambles together when Réaumur visited in the autumn.[13] Baron may well have hoped for an official appointment from the academy as correspondant, but he never achieved this distinction. Still, Réaumur recognized his contributions in print, and the two continued to do favors for one another, Réaumur sending instruments and books from the capital, and Baron sending caterpillars, cocoons, and drawings, along with his letters. This circle of exchange was typical of many of Réaumur's epistolary relationships: the flow of specimens toward Paris warranted requests for return favors of various sorts, even when the personal connection was marked by difference of social status. In due course, the provincial observers would be recognized in print for one or the other of their contributions.

Baron's frustration with his situation gives us some insight into the daily obstacles to the pursuit of natural history, as well as its rewards. "I have not been able to devote myself to natural history as I would have liked, given that I have not had a moment's peace in my practice."[14] This was not just a matter of time, but of timing, as the pace of the insects' life cycles did not always match the doctor's professional schedule, causing him to miss out on events in their development:

> I almost always have the misfortune, when I am ready to send you my little shipment, either to be called to the country or to find that my insects metamorphose [too soon]. I am inconsolable about what has just happened. I had three beautiful caterpillars, one from the *salicavie*, the other from the *lisomache*, and another from a kind of spurge [*titimale*]. I was ready to send them to you when I was called [to a patient], where I was obliged to stay for five days. Upon my return I found one of my caterpillars dead and the others ready to make their chrysalises. Of these two, worms ate one, and the other is worthless. I hope that everything I am sending you arrives safely.[15]

Sometimes Baron had to watch in frustration as the insects he collected either died for lack of food or spun cocoons while he waited for an appropriate conveyance to take them to Paris. "Nevertheless," he wrote, "I am sending you all the ones I have left."[16] He knew that, in order to be useful, he had to devise ways of shipping living and preserved specimens, appropriately labeled and protected from the vagaries of the journey. Initially Réaumur complained that the contents of Baron's shipments had become jumbled in transit, and no longer occupied their assigned niches in the apothecary's box that had seemed well adapted to the purpose. Then Baron was frustrated by not finding enough items sufficiently interesting to send to Paris. He claimed to have spent four months "searching in vain" for something striking or unusual; failing that, he sent anything he could find on the bushes and trees around his house. Even though Baron had prior experience with cultivating insects in Réaumur's home, he often had difficulty matching insects with the appropriate food. Of one of his specimens, an unidentified black caterpillar he found in the house, he noted, "I do not know what it eats. I gave it several plants that it would not touch. It made its cocoon for me and then turned into the butterfly that is in the box." In the same box, he sent caterpillars from a willow tree, a cluster of egg cases, "some worms with blue bellies," and several other live caterpillars.[17]

Another informant and collaborator in the Poitou circle was Charles-René Girard de Villars, a doctor and local notable living near La Rochelle. In the earliest of his surviving letters to Réaumur, from 1732, he refers to his largely unsuccessful efforts to obtain insects from correspondents in the Americas (Canada and Saint Domingue). The only kind he had received was something he called a "Rat-worm," of which he sent a dead specimen. The creatures had still been alive when their box arrived at La Rochelle, but most died soon thereafter, presumably from the cold. He nurtured the two survivors attentively: "I feed them very carefully with sugar, white bread, and some bits of fruit cooked in a little stove, according to instructions from my friend Lacroix."[18] He would have sent one of these living insects if he had not feared its death en route.

Villars kept a great many insects in captivity, and reported on whatever he found locally. In response to a specific inquiry, he noted, "I have not yet been able to discover anything on our broom bushes to indicate that horseflies are born from the spittlebugs. I remember having seen a kind of worm there."[19] Réaumur had apparently suggested that common spittlebugs might be the nymph form of horseflies. When spring came, Villars examined the bushes again: "On the 7th of this month I started to see spittlebugs on our broom. . . . I found on it little insects as big as a flea, more or less; I made out a head, a little pointed nose, etc."[20] He collected bunches of the broom shoots filled with the foam made by the insects, but when the leaves wilted, the foam dried up and "my insects died on the second day." He tried putting freshly cut branches in water, but the insects "languished" and could not make their foam with the proper degree of froth. While he was investigating these spittlebugs, he found on the same plant little white oval cocoons, which he opened to find broken brown shells within.

In the same period, Villars was keeping cockroaches in boxes and watching them reproduce. "Two days ago, in the box where I was carefully keeping my large cockroaches, there were little insects; after an exact analysis, they turned out to be cockroaches. I have put them in with their parents, so that you can see for yourself, sir; their food is bread, sugar, and *cassonade* [unrefined sugar]."[21] He sent a box with specimens of these roaches and the spittlebugs. Villars was quite an assiduous observer and continued to send shipments of various caterpillars, worms, and butterflies in a special two-layered box, divided up so everything would stay in place. He even gave instructions for opening the box, within which each insect occupied its own little compartment: "In order not to spill anything, the white part of the box must be on top; if you [first] take off the white circle, the red section will then be seated in its proper place. In the red part are various live caterpillars with the food that I thought suitable for them."[22] The considerable care and ingenuity devoted to packing his insects betrays the intensity of his engagement with the problem of getting these local finds safely to Paris, where they could take their place alongside the others, from Poitou and elsewhere, in Réaumur's workroom.

BAZIN IN POITOU AND STRASBOURG

One of Réaumur's most extensive surviving correspondences, with Gilles-Augustin Bazin, began in this period as well. Parisian by birth, Bazin trained as a lawyer and was certified as *avocat au parlement*.[23] It is not clear whether he ever practiced law, apart from serving as supervisor of the Paris salt depot from 1711 to 1720.[24] His acquaintance with Réaumur probably dates to this period; they lived in the same Paris neighborhood, and Bazin visited the Charenton country house, where he participated in various kinds of observation.[25] In September 1732, Bazin traveled with Réaumur to Poitou, where he stayed on as a kind of informal agent for more than two years, reporting on estate matters and maintaining a program of collection and observation of insects. At this time, Bazin was nearing the age of fifty; he had just separated from his wife, for reasons that are difficult to determine, and was "taking refuge" in Poitou.[26] His correspondence with Réaumur would continue intensively for more than a decade.

In print, Réaumur described Bazin's contribution thus: "One of my friends, having developed a taste for solitude and for studying our little denizens of the countryside, happily chose my estate for his retreat. There he sought out insects with such attention and patience that even those that seem completely hidden on account of their tiny size could not escape. He enjoyed feeding them, raising them, and he carefully sent me those that seemed to him the most worthy of being followed."[27] Bazin's year-round presence on the provincial estate made it into a prime location for new observations and a rich source of material for Réaumur's work in Paris, where observations could be confirmed or replicated on the specimens arriving by post. When Réaumur came for his annual stay on the estate, he and Bazin worked together, meeting up with Baron and Villars as well.

An adept observer who was also quite skilled at drawing, Bazin kept a steady stream of letters, insects, and drawings flowing from Poitou to Paris. Most of the insects he observed could be found in any garden; his sustained attention and the pains he took to keep them under his gaze, however, transformed them into material worthy of the Academy of Sciences. Bazin's letters, rich in detail and full of enthusiasm, give a vivid sense of how he threw himself into his avocation as a naturalist: "The honor of writing to you without speaking of insects more or less, would be like writing to one's mistress without speaking of love."[28] This affective dimension of the quest for ways of seeing into insect lives was quite common among observers of insects, and it was sustained by the personal attachment of the observers for each other. Bazin certainly enjoyed the thrill of the hunt for hidden treasures like insect eggs, just as he enjoyed the challenge of keeping insects alive and devising ways of watching them; but one cannot imagine him pursuing the challenge without communicating his discoveries to like-minded friends. The bonds of pleasure and sentiment tied the observers to each other as well as to the objects of their study. Reporting on his discoveries, Bazin often reflected on his emotional engagement with the task of finding novelties or correcting misconceptions in the existing literature on insects. As he wrote to his friend and mentor:

> One of the most enjoyable things is finding the eggs of an infinite number of insects. It gives me great pleasure to discover all those little hiding places or to surprise the mothers hiding their treasure. It really is catching nature in the act: because it is so admirable to see the infinite variety of means by which nature has inspired these animals to calm their anxiety and their love for their offspring; this is what I search out with the greatest care, and it is not easy to achieve this goal. For example, this little worm of the ash tree that I found on the figwort, just coming out of its egg; this egg was attached to the newest tip of the plant and seems to have been laid there very recently. It doesn't appear to have emerged with the bud as Swammerdam says it sometimes does. I would really need your eyes, Monsieur, in such a case.[29]

Bazin rarely expressed his admiration for nature's variety so overtly as he does here. He moves deftly from very general admiration, verging on cliché, to the tiny and often elusive details that make up the treasure for which he is searching. The infectious excitement at finding unexpected things in hidden corners of the woods or garden is not enough; he must also take meticulous care to determine how and when the egg was laid on the new shoot and to see the tiny worm hatching from a single egg on the growing tip of the figwort. He also knows what Swammerdam, the reigning authority on insect metamorphosis, has said about it, and points out the discrepancy with his own observation, before deferring to Réaumur's expertise.

While he lived in Réaumur's manor house, Bazin had sufficient leisure to collect, observe, and describe as much as he pleased; he was also dependent, at least temporarily, on the goodwill and even the financial help of his mentor and benefactor.[30] Like Villars and Baron, he sent all sorts of specimens to Paris—larvae, caterpillars, eggs, cocoons, flies, spittlebugs, beetles, and butterflies. Some of these were things he could not identify, some were larvae or eggs Réaumur had specifically asked him to look out for, and some he suspected of being the larva or adult form of an already familiar insect. Identifying things often meant keeping them in containers to watch their transformations; this in turn required knowing, or guessing, what to feed them. Beyond simply identifying and deducing sequences from egg and larva to nymph, chrysalis, and adult, Bazin tried to figure out how the insects laid their eggs, rolled up leaves, spun cocoons, and so on. He was always looking for novel and unusual specimens, and when he found one, he tried to send an example to Réaumur while keeping a few for himself.

The scale of Bazin's enterprise was remarkable—he kept hundreds of bugs, chrysalises, egg cases, and butterflies under observation in jars and other containers. He had several rooms designated for these containers and sometimes moved them into his bedroom for more convenient access when the insects were about to transform or molt. Observing the anatomical peculiarities of a "remarkable" inchworm [*arpenteuse*] with his magnifying glass, he notes that he can find no sign of a mouth and describes various structures whose functions he has not yet understood: "It has now been eight days that it has remained perfectly immobile—and yet it is alive because when I tried to make it bend a little it realigned itself in its original direction."[31] By the next month, he had three inchworms of another species in jars and "supplied them with some dirt as soon as you advised me to do so." These were a "lovely green with a white stripe on each side."[32] Three weeks later, "our inchworms" were becoming chrysalises in the jars: "They did not make use of [the dirt in the containers] but made their chrysalises by hanging from a loose web in which they entangled several twigs of broom."[33] Later in the season, Bazin reported that he had found a string of eggs encircling a branch of broom, which he had brought into the house. He knew the butterfly that had laid the eggs but did not know which caterpillar would hatch from them. "These eggs hatched at the beginning of this month in your back room—which no one would say is warm—and gave me about a hundred little inchworms that I immediately brought into the room where I sleep. In spite of all my care, some of them died every day. I don't know if any of them will survive."[34] Shortly thereafter he was able to report that he could recognize these little worms as "real inchworms," though they still were not flourishing. "Of my whole brood I only have two left, but they are doing very well. One can already see, without a loupe, their headpiece with its two horns."[35]

Bazin's letters follow the development of these caterpillars over many months, as he kept Réaumur abreast of their progress, and the progress of his own understanding. Like the inchworms themselves, this narrative thread wends its way through other ongoing observations, stopping and starting, and coming into focus against the backdrop of a profusion of other insects whenever he has anything new to report. Bazin moved continually from the garden into the surrounding countryside and back to the various rooms where he kept his jars under surveillance and where he wrote his letters. The insects were distributed around the house, with eggs kept in the cold back room and the newly hatched inchworms brought into the relative warmth of the bedroom, where they might be more likely to survive; rows of jars must have been everywhere. The longer Bazin kept the inchworms alive, the more invested he became in their welfare and their transformations. The few survivors of the teeming brood rewarded his attention by pupating and eventually transforming into egg-laying butterflies to bring the process full circle. Although extracting references to the inchworms takes them out of the array of the other creatures competing for Bazin's attention every day, the story of their development over several months was the implicit goal of the sequence of observations. This kind of extraction mimics Réaumur's own procedure as he pulled material out of the letters for his published observations. At the top of each letter from Bazin he jotted notes on key topics and novelties as a rudimentary index to their contents.[36]

Bazin did not follow any single strategy in his collection or observation practices. His work drew on a combination of happenstance, ingenuity, persistence, and perspicacity, as well as excellent eyesight. In a single letter, he might jump from a butterfly caught in the act of laying her eggs, to a metamorphosis taking place on the desk in front of him as he writes, to the contents of the box he is sending, to an update on the condition of specimens reported previously.

In addition to sharp eyes, the naturalist needed the ingenuity to devise new ways of seeing. Wondering how a butterfly manages to break out of its shell, for example, Bazin cut a hole in one end of the chrysalis and attached a piece of glass from a medicine bottle over the opening, so that he would be able to watch the butterfly inside.[37] The first attempt was not successful, but he subsequently perfected the "invention" on several different species, attaching fragments of glass to cocoons with a watery glue. On silkworms it worked perfectly since the cocoon had plenty of empty space, enhancing visibility. A "lawn caterpillar," packed much more tightly into its shell, yielded more startling results. "As soon as the skin of the chrysalis had been opened, I saw the head swimming in a red water. It stayed there for about a quarter of an hour, after which I saw the butterfly strike its head against the wet part." The insect emitted a liquid from its mouth throughout the process, which Bazin watched intently. He was even able to collect some of the liquid, and tasted it: "It seemed to me perfectly insipid."[38]

Bazin's letters give a vivid impression of all the elements of his program of observation: looking, opening egg cases, manipulating chrysalises, wondering about mechanisms, feeding larvae, making drawings, sending material and queries. They also provide some insight into the motivation behind this kind of work—there were always unsettled questions and mysteries and frustrations. "But how could a fly roll up a leaf like this? . . . I will spare no trouble to discover the answer."[39] Things developed continually in the insect world, whether in jars, in transit by the post, or out in the hedges or the fields. Even as he wrote, the objects of his attention might change form: "During the time I was writing this letter," he added in a postscipt, "two of my yellow worms from the rolled-up leaf have just turned into nymphs similar to those I got from that little beetle."[40] Bazin served as an invaluable assistant and collaborator, feeding material into the complex enterprise that underpinned Réaumur's published volumes. Sometimes he followed specific instructions and fulfilled requests, but he was also free to ask his own questions, devise his own techniques, and reap his own rewards from his work. He took pleasure in his observations, and the more adept he got and the more he learned, the more he enjoyed it. His letters paint an exceptionally rich picture of this cycle of work, pleasure, interrogation, and understanding. The pleasures of noticing, gathering, cultivating, drawing, and preserving motivated the pursuit of natural history in the lives of many different kinds of people. To reduce this activity to amusement would be to overlook the intensity of intellectual engagement and the sustained effort and ingenuity directed toward the larger project of puzzling out connections and understanding the dynamics of insect behavior and physiology.

Bazin's situation in Poitou was not permanent. Initially he hoped to acquire property of his own in the area, and he seems to have looked for a post somewhere nearby, but these possibilities did not materialize. In 1733 he got an offer to come to Strasbourg from Jean Vivant, the uncle of his estranged wife. Vivant, who himself had an interest in natural history, had recently moved to Alsace as auxiliary bishop, under the Cardinal de Rohan.[41] It took a year to negotiate the arrangements, but ultimately Bazin moved to Alsace, where he spent the rest of his life.[42] In his new home, he had plenty of time to continue his insect observations, and he cultivated similar interests in his circle of friends and associates. For years, he kept up an active correspondence with Réaumur, sending live and preserved specimens, eggs, drawings, and books to Paris. In 1737 he was rewarded with the status of official correspondent of the Academy of Sciences.

Bazin continued to perfect his skills after settling in Strasbourg, adding dissection to the list of techniques he routinely applied to insects. He had been dissecting caterpillars for several years when he decided to attend a course of lectures on human anatomy at the university. There he learned inflation and injection techniques commonly used in anatomy demonstrations, which he adapted for use on insect vessels to make them more visible.[43] By exposing the internal structures of caterpillars and butterflies, Bazin discovered certain peculiarites of their physiology, noticing many details Réaumur had not seen. He maintained quite an extensive operation, keeping numerous insects in various stages in a study where he could also dissect and inject specimens for preservation.[44] Unlike many of Réaumur's correspondents, Bazin did not complain of restrictions on the time he could devote to his insects, though he certainly had other duties. The bishop, his patron and employer, encouraged his investigations; the observations and dissections provided a point of interest for neighbors and visitors, who stopped by to admire a caterpillar's circulation or the action of its jaws. As he had done in Poitou, Bazin integrated his science into his domestic space, adapting the study and the furniture to his very particular ends.

As he developed his dissection techniques, along with many other experimental protocols, Bazin became less dependent on Réaumur's direction and suggestions, although he did ask for advice from time to time and continued to send packages to Paris. While working on a given problem, Bazin often sent duplicate caterpillars or chrysalises to Réaumur, so he would be able to replicate observations in Paris before they turned into butterflies. He maintained a long-distance dialogue with his mentor, reacting to suggestions and providing new observations to enrich the academy and Réaumur's future volumes. The shipments allowed a kind of shared experience not otherwise possible at such a distance. Sometimes Bazin sent drawings, but he preferred to send live insects whenever possible. At other times, he sent anatomical preparations as a way of displaying the results of his minute dissections. In one such instance, he could not help boasting a bit: "I am so excited by my success, that I would be mortally disappointed if you did not witness it. I have done the impossible, as you wished, sir; I have detached the heart of a caterpillar from its surrounding tissues."[45] To document his achievement, he sent an injected preparation of the dissected caterpillar pinned to a plaque, though he was not sure it would survive the journey.[46] Réaumur reciprocated with lenses, thermometers, and books; he also transmitted letters and financial documents to Bazin's brother or his lawyer in Paris. When Bazin and Vivant found themselves wishing for quality chocolate, not readily available in Strasbourg, Bazin asked Réaumur to acquire some cocoa paste from a merchant he knew in Nantes—which he did willingly. In return, in addition to sending insects (including a mite that was fond of chocolate), Bazin served as a convenient conduit for books and letters sent to Réaumur from Germany, notably from Christian Wolff and Albrecht von Haller. So along with their lively correspondence, which covered personal matters as well as natural history, the two men were continually exchanging other favors and objects. Like many savants around Europe, they did not calculate the worth of these exchanges on a quid pro quo basis, but rather took them for granted as the routine apparatus that supported long-distance friendship and collaboration.

Bazin's letters are so packed with the details of his many investigations that it would be impossible to follow all of them here. He filled page after page with his minute script, in a kind of insectological logorrhea: "I feel completely full of wonderful things to tell you. I advise you, however, to put off reading this until you are in some inn where you find yourself with nothing to do."[47] In one instance, he wrote excitedly of his discovery of a caterpillar's tongue: "I spent the entire morning yesterday looking at the tongue of this animal, which is clearly visible. . . . I will only be fully confident of the evidence of my eyes when yours have confirmed my report."[48] He sent along some healthy caterpillars, in case Réaumur wanted to attempt the vivisection himself. I quote at length to give the flavor of Bazin's homely practice, described vividly here:

> I have an armchair with deep plush upholstery; when one is kneeling [on the seat], one can rest one's elbows comfortably on the wide top surface of the chair's back. Down below this, on the rear surface of the upholstery, I attached two strands of thread, separated from each other by the length of a medium-sized

> caterpillar. I bring the threads up over the top of the armchair; each one has a loop on the end which comes to the front. I set my caterpillar on the top upholstered surface, I pass one of my threads over its neck, and, with a pin that I insert through the loop of thread and then stick into the upholstery, I immobilize and strangle my caterpillar tightly. I do the same with the other thread around the tail end. You can easily see that immobilizing the animal thus at both ends leaves the scalpel free to operate on its body. Having thus adjusted my caterpillar, I examined all the parts of its body before opening it. Noticing that this strangling caused it to open its mouth very wide, I immediately took my magnifying lens and saw it make all the motions it makes when it eats. [49]

After watching this spectacle for a time, Bazin sliced into the body with his scalpel. At this point, some of the contents of the intestines fell onto the tongue: "I then had the pleasure of seeing the play of the tongue as it pushed the particles of food toward the esophagus."[50]

Evidently the pursuit of natural history had taken root in the bishop's residence in Strasbourg. "You would certainly find it edifying, sir, if you saw me with lancet, knife [*bistouri*], and tiny scissors dissecting the head of an insect; you would say that this man has a passion for anatomy. I am on the eighth head, where I see as clearly as in that of an elephant; this is not the case for the body, where the parts are still very confused for me."[51] Bazin demonstrated his findings to neighbors and visitors, developing a reputation for his unusual skills. A local doctor admired the "skill and delicacy" of his dissection technique when Bazin was able to show him the membrane that separates the caterpillar's esophagus from its stomach.[52] Sometimes he invited friends in to confirm unexpected findings. In the case of a large butterfly he was dissecting, "After having opened and firmly attached this female [butterfly], I had the pleasure of watching it continue to lay its eggs, even though it was pinned down. Then I turned to the heart which was of considerable size. But my greatest surprise was to see the circulation moving in the same direction as in the caterpillar, that is to say from bottom to top." Though he could see the motion clearly, he asked two friends to confirm what he was seeing.[53]

Bazin provided as much detail about his experimental techniques and setups as about the surprising or unexpected phenomena he observed in his dissected insects. Even when he reached no definitive conclusion, he might report on his maneuvers and inventions to show off his novel methods for seeing into the bodies and the lives of caterpillars and butterflies. The correspondence documents much more than a series of discoveries, or particular claims about anatomy or physiology—although Réaumur was happy to have these, to be sure. As much as he was revealing the inner workings of his insects, Bazin was

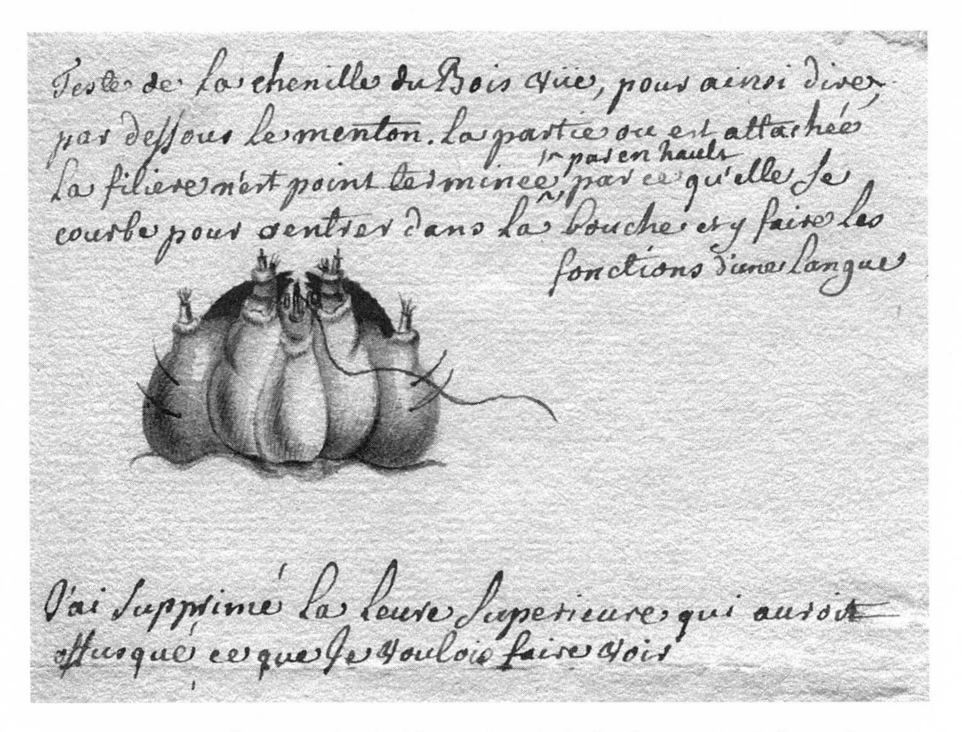

Figure 2.2. Gilles Bazin, sketch of dissected caterpillar head, 1735: "Head of a wood caterpillar viewed, so to speak, from underneath the jaw." Archives, Académie des sciences, Fonds Réaumur, dossier 42. By permission of Archives de l'Académie des sciences, Institut de France.

✲

also demonstrating his own dexterity and ingenuity. He submerged caterpillars in liquid to see where air bubbles appeared. He dipped butterflies in oil to keep the powdery particles on their wings from irritating his eyes during operations, and was astonished to see that the oil made the body transparent. "Scratching the underside of the body with a penknife, you see everything going on inside."[54] Or the heart of an apparently dead insect might start beating again when he started the dissection. Part of the game was to see how far he could push his dissections and vivisections, adapting his setups to address physiological questions about respiration, circulation, and metamorphosis. As one last example, here is his description of experiments designed to determine whether caterpillar blood circulated, and if so, whether it delivered air to the tissues:

> After preparing a caterpillar on a pierced plaque of wax, as you know I customarily do now, I put this plaque in a crystal saucer and put this saucer on something

> black. . . . I poured water into my saucer until the caterpillar was covered. Then passing my lancet through the water, I pierced the heart, thinking that if there was air inside, it would come out in bubbles, and rise to the surface of the water. In vain I pricked and tore that poor heart—it did not yield even the least appearance of air.[55]

He tried again, cutting crosswise through the animal, to see if any liquid might be escaping from the heart; the black background under the saucer should have made such movement visible, but he saw nothing through his magnifying lens. He finally decided that the analog of blood in caterpillars was nothing more than "the fluid that comes out of their body when one wounds them somewhere"; instead of circulating, this liquid "spreads throughout the inside of their bodies."[56] Bazin then investigated the properties of this fluid, testing its reactivity with various chemicals.

THE PINE CATERPILLARS OF BORDEAUX

Bazin was part of the sociable circle around Réaumur in Paris before leaving the capital for the provinces, and these personal ties facilitated their extensive exchanges over the years. Other epistolary relationships started up when distant naturalists or "*curieux*" contacted Réaumur out of the blue and ended up supplying valuable knowledge and specimens. Many of these collaborators and correspondents can be spotted in the pages of the *Mémoires pour servir à l'histoire des insectes.* One such person, who played a key role in Réaumur's investigations of the pine caterpillars native to southern France, was a magistrate in Bordeaux by the name of Raoul. He lived in town but also owned a country estate with a considerable stand of pine trees.[57] As a landowner, he was concerned with forest management and with the pests and diseases afflicting grain and other crops. He took an interest in the insects of his region, collecting local knowledge from peasants, apothecaries, and other landowners. At first glance, he appears to be a model of the enlightened men associated with provincial academies, but in fact Raoul had only scorn for the notables who gathered at the Bordeaux Academy, which he dismissed as an academy "in name only." "If I discover something definite and useful, I will be honored to communicate it to you rather than to our academic do-nothings, who have not yet even discovered the secret of getting rid of the moles and caterpillars that, by the way, have devoured all our apple trees."[58] Raoul kept up with the literature emanating from Paris; he seems to have read the *Mémoires* of the Paris Academy as soon as they appeared, and he was familiar with Réaumur's recent papers on clothes moths.[59]

Raoul had lived in Paris for a time in his youth in the 1690s, but he did not know anyone there who could serve as an intermediary with Réaumur.[60] When he decided to contact the academician to report on experiments designed to destroy the weevils infesting his wheat stores, he masqueraded as a village priest and signed his letters with a pseudonym, "Pradat, curate of Begadan."[61] Later, when he sheepishly dropped this ruse, he explained, "Not having the honor of being known to you and not knowing personally the extent of your politeness, I thought I should hide myself until I saw that you did not scorn the observations of a Gascon."[62] He could not have known that Réaumur was happy to correspond with virtually anybody who had done experiments on insects and who might supply novel material.

In his first (pseudonymous) letter, Raoul/Pradat explained how he had mixed salt into some of his stored grain to see if it would kill the weevils. This provincial landowner was completely conversant with the language and methodology of experiment, keeping grain in separate boxes under different conditions to test his idea.[63] He monitored his containers systematically, watching to see if the insect eggs hatched in the treated grain. Unfortunately, though salt worked well as an insecticide, it ruined the grain by attracting moisture and rot, and when ground into flour it produced inedible bread. The next year he tried another remedy sent to him from Angoulême, painting the walls of his storage barn with a solution of white vitriol: "I will make the test as soon as I am in the country. My apothecary maintains that vitriol does not absorb water in humid weather."[64] Raoul also mentioned some peculiar insect nests he had noticed in his pine trees, made of "a very strong and very white silk."[65] Réaumur encouraged him to continue his experiments with the wheat pests and asked for some live pine caterpillars, which he had not seen in Paris. Raoul obliged by sending a package, and continued to send shipments over the next several seasons. Over time, he got more and more expert at observing the pine caterpillars himself, following them through their life cycle and studying the properties of their silk. Réaumur's desire for a continuing supply of caterpillar nests and eggs, and his probing queries about the habits of this unfamiliar insect, inspired his provincial correspondent to pay more attention to the caterpillars, bringing them into his house and finding ways to keep track of their movements.

In August 1732, Raoul searched unsuccessfully at his country estate for eggs in the sacs spun by the pine caterpillars, finding only their shed skins. He promised to try to discover at what season they transformed into butterflies, and how and where they laid their eggs. In May 1733 he reported that he had successfully charted the development of the caterpillars since the previous fall, up until the time they changed into chrysalises. "I would have followed them to the butterfly stage, but without my realizing it, my servant threw them away while sweeping out my study, and I only noticed it four days afterward."[66] The insects made large nests of pine needles bound together with remarkably strong white silk. He then described what he had seen from the time of their hatching at the end of October, when their fine threads, the beginnings of the nest, first became visible on the pine branches. He was able to see the caterpillars leave the nest each day, through an opening in the bottom of the nest. "They follow each other to where they want to feed and leave a trace like a strand of silk or ribbon a half-line in width. When they have eaten, they return by the same path into their nest, which they enlarge and augment bit by bit. When the weather is cold or rainy they do not go out. They leave toward sunrise and return two hours later, or thereabouts, to their nest, which is impermeable to rain."[67] This report seems to have been based partly on Raoul's watching nests in the trees, since he refers to their behavior in bad weather, and partly on watching the residents of a captive nest. He kept at least one nest in his study: they left a silky trail across the chimneypiece. "I hung a bunch [of pine needles] from [the mantel] on a long string; in climbing up and down they covered it with silk." Even after the servant had carelessly discarded the insects, the traces of silk remained in his study. The following winter, he sent more caterpillars to Paris, along with a quantity of silk separated from the pine needles of the nest. "I will have several pounds of it collected and clean it as well as I can and send it to you. It will be curious to have some stockings made from it, which I think will be durable, since it makes quite a strong thread."[68] It seems likely that Raoul knew of Réaumur's investigation of spider silk, and the story of the stockings knitted in Montpellier (see chapter 1).

Raoul was not the only one in Bordeaux curious about these insects. In 1734 his friend Dr. Cardose asked for one of the nests. He took it to show to the Bordeaux Academy and then kept it in a box with some pine needles. The caterpillars did not come out of the nest to eat, however, for seven weeks. "Finally yesterday we visited them," Raoul reported in April. "We found a lot of fine threads that crisscrossed the box, covering it. But what will surprise you is that some of them left their nest and formed in several places in the box about a dozen little cocoons where they enclosed themselves, like those made by silkworms but much smaller and with a very white silk. They seem to be still working on enclosing themselves. Thus, sir, I hope to send them to you in eight or ten days. I will not fail to do the experiment as you wish, and to repeat that of Mr. Cardose."[69]

In his chapter on pine caterpillars, Réaumur described his own program of observations in Paris, made possible by Raoul's shipments of nests.Read in tandem, the letters from Bordeaux and the published account chart the two investigations as they proceeded in dialogue for several seasons, in Bordeaux and in Paris, with letters passing back and forth, along with nests, eggs, and caterpillars. "The caterpillars that inhabited [the nest] seemed to have suffered not at all from traveling by the post," Réaumur noted.[70] The caterpillars appear in Réaumur's chapter on social insects, though they first aroused his curiosity because of the properties of their silk.[71] He had trouble for several seasons keeping them alive until they transformed and laid their eggs. The strength of the silk seemed to indicate that it could perhaps be carded and spun. "M. Raoul, to surprise me with a courtesy of the kind he thought I would most appreciate, had a lot of this silk collected with the aim of having it prepared and having a pair of stockings made for me; a lady took on this task."[72] But when the lady had the silk boiled with soap to clean it, the fibers disintegrated. Réaumur repeated the experiment, "on quite a small quantity of this silk," and discovered that as soon as it was put into boiling water, it broke up and seemed to dissolve. "This invites us to do new experiments, to see if there is in nature a silk that boiling water can dissolve. Such a silk would perhaps be useful for the construction of flexible varnishes and unusual fabrics."[73]

Though he never got his stockings, Réaumur did examine the construction of the nests closely and had drawings made of the structures of pine needles held together with caterpillar silk. Each nest was built by the caterpillars hatched from the eggs of a single butterfly. After the failure with the first nest from Bordeaux, he took the next one to the Jardin du roi, where his friend Bernard de Jussieu agreed to hang it from a pine tree kept in a greenhouse; the caterpillars didn't survive long in this environment. In March, Raoul sent another nest, and this time Jussieu hung it in a pine tree growing outdoors. After two days, all of the caterpillars had disappeared. "Luckily I had kept about twenty others in a big jar [about half full of dirt]; I gave them pine needles [to eat], which they didn't even seem to notice; it seemed to me that they didn't touch them."[74] Finally, all of these caterpillars disappeared too, though now it was obvious that they must have retreated into the soil in the bottom of the jar. A few weeks later, he found their buried chrysalises, and then dug up some others under the pine tree in the botanical garden. Eventually he saw the butterflies emerge from the chrysalises, completing the cycle.

Raoul's observations paralleled those unfolding in Paris. Without dirt in their box, his caterpillars could not bury themselves before they wove their cocoons, so they were easier to find. In April he sent a box of the cocoons, but he could find no more caterpillars in the forest: "You cannot imagine how many nests I cut open without finding anything."[75] When he wrote all this up, Réaumur presented the caterpillars' growth and development as part of the story of his

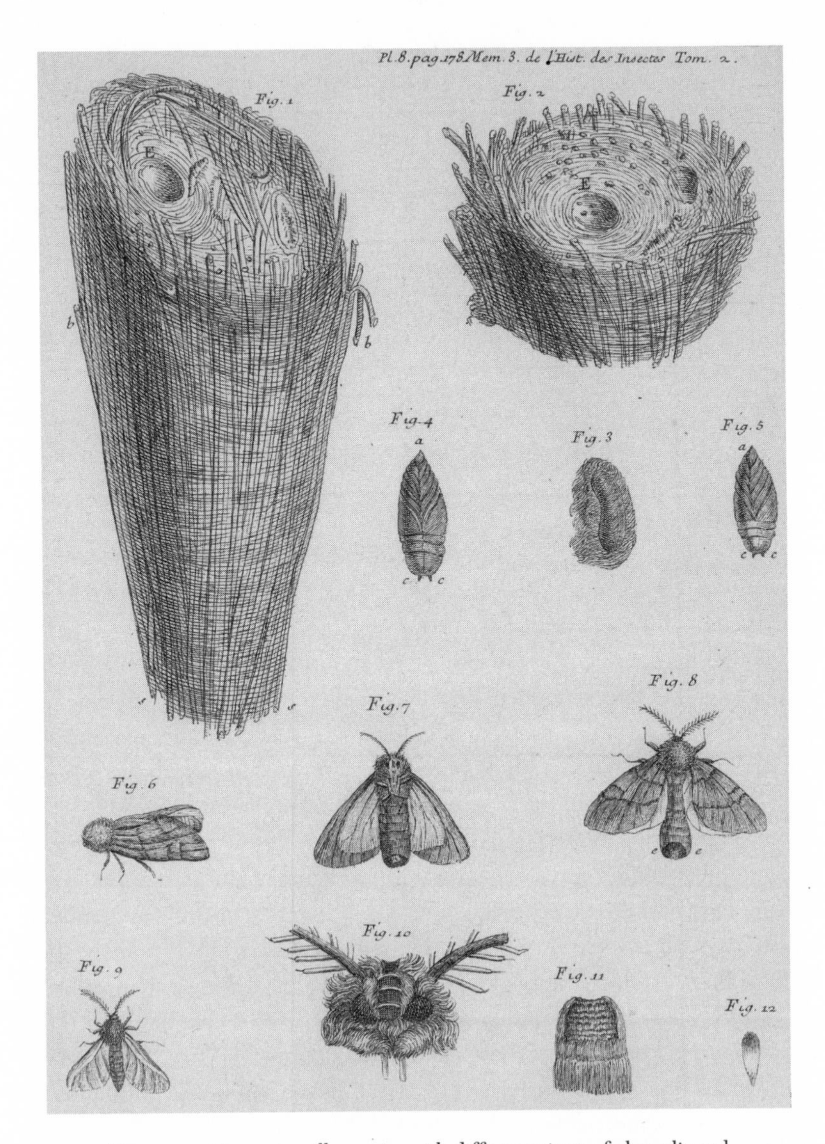

Figure 2.3. Pine caterpillar nests, with different views of chrysalis and male and female moths. Engraved by Philippe Simonneau from drawing by Hélène Dumoustier. Réaumur, *Mémoires pour servir à l'histoire des insectes*, vol. 2, plate 8. Courtesy of History and Special Collections Division, Louise M. Darling Biomedical Library, UCLA.

own investigation, including the crucial role played by Raoul. Along the way, he drew attention to his setbacks and successes, as object lessons for the reader, who could see the intimate connection between the observations of captive insects and those left outdoors where they escaped notice, for a while, in the earth. The detailed report on the pine caterpillar reads as a tale of discovery, with the investigators resolving apparent conundrums one after the other. In the final account of the life cycle, habits, and anatomy of the insects, the actions and decisions of the observers have become part of the story. In typical fashion, the entire sequence of observations required moving back and forth between the insects in the jar and those in the garden, which for Réaumur meant traveling across town to visit the pine trees, and allowed him to extrapolate from the behavior of the captive insects to those living outdoors.

The people and insects we have encountered in this chapter represent only a small fraction of the similar stories that could be told about natural history in this period.[76] Through these examples we have seen a variety of people in different parts of France—Poitou, Bordeaux, and Strasbourg, as well as Paris—engaging with nature and with each other. Each of the main characters of this chapter had a circle of interlocutors who visited each other, went on collecting excursions together, and shared specimens, observations, and opinions. Following insects—catching them, feeding them, watching them, cutting them open, packing and shipping them—was mundane work, integrated into the routine and the spaces of domestic life. In Bordeaux, Raoul's pine caterpillars lived in his study (until the servant discarded them), and boxes of weevils submitted to the salt treatment in his barn. The provincial doctor Baron tried to keep his collections alive in his modest study, with only partial success, while the more prosperous landowner Villars fed a population of cockroaches on sugar and white bread in his house nearby. The fields and woods around the manor house in Réaumur supplied Bazin with the many species that filled jars and boxes in various rooms and outbuildings; when he moved to Strasbourg, the bishop's household played host to a similar range of insects. In the following chapter we return to Paris, and the way natural history was integrated into Réaumur's own household.

3

Seeing Again and Again

ILLUSTRATION AND OBSERVATION IN DOMESTIC SURROUNDINGS

Anyone trying to follow insects through their maneuvers and transformations had to incorporate these creatures, one way or another, into their domestic routines and spaces. This might mean, as it did for Bazin in Strasbourg, adapting the back of an upholstered chair as the stage for caterpillar dissections; or one might give over one's chimneypiece to pine caterpillars and their nests, as Raoul did in Bordeaux. In the previous chapter we saw something of the variety of the inventions and procedures that made natural history possible. This chapter looks more closely at Réaumur's several residences as the architectural and social settings for the scientific pursuits of the people living and working there. The composition of the household fluctuated, as did the focus of the efforts and attention of its residents and visitors. Across the intervening centuries, we can detect the activities of the people in these houses only sporadically, subject to the vagaries of the sources. Identifying the main people and situating them in their surroundings and in their relationships to each other will show how natural history worked as a domestic enterprise.

After moving to the capital in 1703 as a young man of twenty, Réaumur made his primary home in the city for the rest of his life, changing his Paris abode several times. For many years he also leased a spacious country house in the riverside village of Charenton, a pleasant ride from the center of the city. With its gardens, sloping down to the banks of the Marne, and ready access to nearby forests and river islands, the house was the perfect setting for doing natural history. The house served as a retreat from the intensity of city life, a place to entertain guests and to do experiments and make observations. In spring and summer, Réaumur customarily lived there for days and even

Figure 3.1. Idealized view of Réaumur's insect "menagerie," where hundreds of insects were raised for observation. Note ranks of glass jars on right, caterpillars and chrysalises (hanging from rack) on table in foreground; ant lion nests, beehives, spider web, and pond with dragonflies outside. Drawn and engraved by Philippe Simonneau. Réaumur, *Mémoires pour servir à l'histoire des insectes*, vol. 1, vignette.

✲

weeks at a time, traveling into town for meetings of the academy and other engagements.[1] He gave up this rural retreat in 1740, when he consolidated his scientific activities, including his expanding collection of preserved specimens, into a comfortable establishment with extensive gardens and yards, located just outside the city walls in the Faubourg Saint-Antoine. (This house will be explored more fully in chapter 6.)

THE HÔTEL D'UZÈS

From 1728 to 1736, Réaumur occupied a lavish *hôtel particulier* a few minutes' walk from the academy's meeting room in the old Louvre palace. This house, in the elegant residential district surrounding the Palais royal, was none other than the Hôtel de Rambouillet, built for the Marquise de Rambouillet and famous for her "blue room," where the *précieuse* hostess gathered the literary and social elite of France in the 1660s and 1670s. In Réaumur's day, the house was known by the name of its current owner, the Comte d'Uzès, great-grandson of the literary marquise.[2] Like most aristocratic residences in Paris, the Hôtel d'Uzès encompassed several wings and auxiliary buildings of various sizes, arranged around two courtyards, with a garden, stables, and cellars.

These spaces proved adaptable to all sorts of uses.[3] There was plenty of room for Réaumur's expanding collection of fossils, minerals, and insects, as well as an extensive library and workrooms for scientific instruments and the insect "menagerie." There was also space for assistants and artists, some of whom lived on the premises.

Even before he took up occupancy, Réaumur undoubtedly knew the place well, through a connection that will take us on a short detour into his work on metallurgy. Attentive readers of the *Mercure de France* for December 1726 would have noticed an article about a commercial establishment recently opened in the Hôtel d'Uzès:

> For several months the Royal Manufacture of Orléans, established at Cosne to convert iron to steel and to make objects of cast iron and steel following the principles of M. de Réaumur, has had a shop in Paris, in the rue St. Thomas-du-Louvre, at the Hôtel d'Uzès. It has been visited by the most distinguished people and the greatest connoisseurs at court and in Paris. Everyone has been struck by the beauty and tastefulness of the objects to be found there.[4]

The genteel audience at the academy's public session the previous spring had heard Réaumur's announcement about this merchandise. "A manufacturing concern, already well established, is working daily to vary and multiply these items. The attention and intelligence of those who have undertaken this, their disposition to regard their interests as inseparable from those of the public, and . . . the considerable funds they have invested, seem to assure the state of our art."[5] The art of casting finely detailed objects of iron came directly out of Réaumur's metallurgical experiments, first published in a nicely illustrated book in 1722 and elaborated over the next four years in additional experiments in the furnaces at the Charenton house. The smelting process produced a white iron soft enough to be dressed and polished after casting, and the finished objects mimicked cast work in bronze or copper, at a fraction of the cost.[6] The Duke of Orléans, then regent of France, rewarded the book's author with a generous royal pension, and shortly thereafter an association of investors obtained a privilege to produce and market goods using Réaumur's "secrets."[7]

Along with his close friend Pierre Jarosson, who also had roots in the Poitou region, Réaumur was actively involved in putting his metallurgical innovations into practice.[8] For years, the two men had shared lodgings in Paris and their families were closely acquainted. Jarosson, a prosperous barrister, invested in the manufacturing company and even traveled to the forge at Cosne, in the Loire valley, to supervise the early phases of production.[9] Réaumur himself spent considerable time at the forge as well, experimenting with modifications of various formulas and procedures. After three years of trials and substantial capital investment, the foundry was producing an array of decorative locks, door knockers, balcony and stairway railings, chandeliers, and wall sconces, as well as various kinds of steel. The Hôtel d'Uzès housed not only the shop, frequented by Parisian connoisseurs, but also furnaces and workshops, where artisans filed, polished, and varnished the cast objects, and cellars for storing steel bars for the wholesale market.[10] The manufacturing and retail business apparently lasted only a few years, perhaps because of complaints from the locksmiths' guild about unfair competition.[11]

Réaumur may have taken over the lease on the house when the showroom and workshops were dismantled; he presented his last experiments on iron to the Paris Academy in 1727, and there is no trace of the ironwork shop after that. Once he moved into the Hôtel d'Uzès in 1728, it became a hotbed of scientific activity. Collaborators, assistants, illustrators, and visitors came and went, engaged in a variety of overlapping missions. Many of these people also traveled regularly to the country house in Charenton, with its fully equipped laboratory and workshop. Indeed, the country laboratory had been the setting for Réaumur's experimental work for decades; with the move to larger quarters in town, it came to function as a kind of annex to the workrooms, collection cases, and garden in Paris. These domestic locations were also linked to the academy, of course. Scientific results made their way from the laboratory and study to the meeting room of the academy, and from there into the pages of the academy's journal and other publications. Colleagues visited Réaumur at home on academic business, and he structured his work around the schedule of semiweekly meetings at the Louvre. Several of the assistants who worked with him over the years themselves ended up, in due course, as academicians.

Henri Pitot, whose association with Réaumur spanned the years when the household was growing, was the first of many such protégés. Like most of the others who worked in the laboratory and collections over the years, Pitot became a friend and close associate, well acquainted with the other members of the household.[12] No set rules governed this loose patronage or mentoring relationship. The arrangement, repeated a number of times over the years, was flexible enough to accommodate a variety of career aspirations, family backgrounds, prior training, and intellectual interests. Réaumur's administrative duties in the early 1720s included supervising the academy's chemical laboratory. Since the institution did not have space for chemical equipment in its quarters at the Louvre, it seems plausible that by the 1720s the laboratory in Réaumur's house at Charenton effectively functioned as the academy's laboratory.[13] In any case, Réaumur appointed Pitot, an energetic young mathematician, as assistant in the chemistry laboratory in 1723 and sponsored him as an "adjunct academician"—the first rung on the institutional ladder—the following year.[14] Though paid with academy funds, Pitot worked directly for Réaumur, first on metallurgy and then on the properties of other materials, like glass and porcelain, and eventually on thermometers. In 1724 he traveled with his patron to Cosne, where they worked together on the new iron-casting process at the forge. As the metallurgical experiments wound down, Réaumur enlisted Pitot's help on his next physics project, measuring the expansibility and compressibility of various liquids at different temperatures and experimenting with procedures for making and graduating thermometers.[15]

Thermometers, though simple in principle, were notoriously inconsistent from one to the other in this period, making comparisons between instruments virtually meaningless.[16] "Someone who has followed a thermometer over several years, who has noted how high the liquid went on the hottest days and how low it dropped on the coldest days, will see in the following years when the heat or cold will be above or below the levels he observed in previous years; but his observations are only for himself, as he will be unable to compare them with measurements made in other places."[17] Réaumur sought to find a ground for comparison among thermometers by tying his scale to natural constants. Speaking to the public session of the academy in 1730 about new "rules for constructing thermometers with comparable degrees," he explained that his thermometer measured "physical degrees of heat and cold." He fixed the zero point of his thermometric scale at the temperature of freezing water, and attempted to standardize performance by filling the glass tubes with a uniform spirit of wine according to specialized protocols.[18] The key was to fill the instrument with an exact number of precisely measured units of liquid, and then to calibrate it by reference to freezing water (or, later, melting ice). In extensive trials conducted with Pitot, Réaumur had devised techniques for graduating the scale, introducing the liquid into the glass bulb, purging it of air, and sealing the tubes.[19]

Generally, Pitot remained invisible and unnamed in print, but readers could spot the assistant at work in Réaumur's narrative of a spin-off discovery a few years after that first report:

> The new construction of thermometers . . . where everything depends on very exact measurements, put me in the way of . . . seeing what happens to the volumes of spirit of wine and of water when they are mixed together. I was thinking only about the filling of thermometers when M. Pitot, after pouring a certain number of measures of water into a glass ball attached to a tube, poured

> onto it a number of measures of rectified spirit of wine, which should have risen up to a certain height in the tube.

The height of the column was duly marked, but Pitot noticed that the alcohol was sitting on top of the water, rather than mixing with it. After shaking the thermometer, "M. Pitot pointed out that our liquid was below the mark, that he would have to add some to get it to that level."[20] In this vignette of laboratory life, part of a much longer history of thermometry and the physics of chemical mixtures, we catch sight, however briefly, of the two men working together in the country house laboratory.

Pitot made a number of thermometers according to what became known as "Réaumur's principles." Several large wall thermometers constructed according to the new method were soon hanging in the homes of wealthy aristocrats and instrument collectors like the Duc de Chaulnes and Pajot d'Onsenbray.[21] Another was mounted outside a ground-floor window on the north side of the Hôtel d'Uzès, and a similar one was installed at Charenton. These instruments imposed a regular rhythm of weather observation on the household for many years; regular temperature readings structured the day, though of course this was only one of many household uses for the instruments.[22] In less than a decade such thermometers had become ubiquitous and were being sent around the world with travelers enlisted to record daily temperature measurements for comparison with those kept by Réaumur himself in Paris and Charenton.[23]

Making instruments and taking temperature readings accounted for only a small part of the scientific activity in the Hôtel d'Uzès and the Charenton villa. In the early 1730s, Dr. Baron, the medical man we encountered in the previous chapter, lodged with Réaumur and took charge of the live insects and no doubt helped out with related tasks. Later, practicing as a provincial physician, he looked back nostagically on this time and wished that he had been offered Pitot's place when the latter finally got a pensioned slot at the academy. "If I had known that Mr. Pitot was going to have a pension so soon, I would have taken the liberty to ask you for the position in the laboratory."[24] In fact, Jean-Antoine Nollet had already taken over Pitot's duties. By this time, the laboratory was no longer tied directly to the academy. As Grandjean de Fouchy recalled later, "M. de Réaumur entrusted his laboratory to [Nollet], and it was in this excellent school, which has supplied the academy with several of its most illustrious members, that he completed his training."[25] Nollet was known for his dexterity and craftsmanship, and especially for his skill at enameling, wood turning, glassblowing, and metalworking, as well as his work assisting Charles Dufay on electrical experiments.[26] He was soon making, testing, and improving thermometers with Réaumur.[27] He also recalled their collaboration on refinements to the thermometer protocol: "We subsequently recognized, he and I, that it was easier and more reliable to take this degree from crushed ice that was just melting."[28]

Nollet eventually parlayed his skills into a profitable career as a physics lecturer and scientific instrument maker. He was still working for Réaumur when he offered his first course of public lectures at his home in Rue du Mouton. A catalog of instruments available from Nollet's workshop, published in 1738, listed five different kinds of thermometer "built according to Réaumur's principles," including large wall thermometers and several smaller, portable versions.[29] Nollet's success as a maker of scientific instruments and demonstration apparatus followed seamlessly from his experience with the development and dissemination of the thermometer. The work of the laboratory at Charenton was not so specialized as this might suggest, however. The range of activities Nollet pursued under Réaumur's supervision shows how natural history, physics, and chemistry fed into each other, and spilled out of the laboratory into other domestic spaces. Though he resided elsewhere, Nollet worked in and around both the town and the country house, handling insects as well as instruments. His duties extended to showing visitors around the collections. The Provençal botanist and antiquary Jean-François Séguier, on an extended visit to Paris, was taken on a tour of the cabinets in the Hôtel d'Uzès by Nollet, "who knew the collections very well."[30] Nollet was standing in for the proprietor on the day of Séguier's visit, indicating the extent to which he was integrated into the household. "I saw there," Séguier noted, " a great number of insects and butterflies preserved between two layers of glass glued together at the edges, and the animals were prepared in such a way as to protect them from the threat of mites."[31]

Réaumur recognized Nollet's aptitude for the kind of close observation necessary to the naturalist, an aptitude honed through his attention to the maneuvers of the insects filling boxes and bottles throughout the house: "While attending to the insects in my little menageries, [he] developed a great taste for observing them; [he] is very talented at discovering even those that are very well hidden."[32] Nollet noticed a particularly small and reclusive aquatic caterpillar under the leaves of duckweed in the pond at Charenton. He brought two of them, enclosed in a little mass of duckweed leaves, indoors for further study. The clump of leaves turned out to be a shell formed of white silk, covered all over with the tiny leaves to make it invisible on the surface of the water. Nollet's encounters with insects were not limited to those he came across in the garden. A rare surviving list of assignments for Nollet, written in Réaumur's hand, includes, alongside measurements of the temperatures of water mixed with various salts, and the quantity of air released from mixing water and spirit of wine, instructions for "suffocating insects coated in butter in the pneumatic machine to see where the air comes out of their bodies." In addition to thermometers and other instruments, he was to construct a special box with different compartments, designed to test the food preferences of wheat weevils, the same small beetles studied by Raoul in the same period.[33] Nollet would have carried out these tasks in Charenton; some of the weevils were brought from town out to the laboratory, where they seemed to thrive in the humid conditions.[34] In the reverse direction, Nollet's aquatic caterpillars traveled back with him to Paris, where they were kept alive in the insect menagerie.

The rhythm of experiments and observations depended on frequent traffic between the two residences, a short carriage or horseback ride; assistants and servants could go back and forth on foot. One of these was a Monsieur Perreau, a young man training to be an engineer and in the meantime employed in collecting and taking care of the insects. Walking to Charenton, Perreau noticed that the dirt in a long stone wall bordering the road was peppered with holes made by a certain kind of bee: "Having dug into the dirt, he found and brought to me the nests that they had made there."[35] After Perreau had brought this insect habitat to his attention, Réaumur kept his eye on bees living in the wall, conveniently located on his route to and from the country house. He transported a few of the bees, and their nests, back home, where they took their places on the worktables, and eventually in text and illustrations, having been examined by numerous eyes and instruments.

These movements of instruments, insects, and people reveal the outlines of a practice of natural history embedded in an experiment-based program using the tools and techniques of the physical sciences. Wall-mounted thermometers registered daily highs and lows of temperature, as we have seen; smaller devices show up not only in the laboratory but in the insect room as well. Thermometry, including experiments on the freezing point of various solutions and the properties of the alcohol used in the tubes, proceeded in parallel with experiments on controlling insect infestations in grain and cloth. Nollet used some of the ice brought in for calibrating thermometers to test whether freezing temperatures would kill wheat weevils. The pests survived the cold but fared rather worse when subjected to heat, measured with the same thermometers. "I had instructed the Abbé Nollet to determine the degree of heat capable of destroying them," Réaumur noted, "so that we might see if we could not successfully use heat against them, and he found that a temperature of 40 degrees causes them to die in less than four or five minutes."[36] This led them to wonder if a similar degree of heat might destroy the larvae of clothes moths (*teignes*), a common household pest and a subject Réaumur had already discussed at length in the academy.[37]

In the same period, when Nollet was heating and freezing weevils and populations of clothes moths were captive in jars eating their scraps of wool fabric, Réaumur was also experimenting with the effect of temperature on the development of chrysalises that would normally hibernate in the cold months. In January, to find a place that would be warm around the clock, he took a selection of chrysalises from his house across the city to the new greenhouses his friend Dufay had recently built at the Jardin du roi. Sure enough, within a few weeks, he had butterflies that would otherwise have hatched in May, or even later.[38] While he was shuttling back and forth to the greenhouse to check on his insects, the inspiration struck him that he had another source of constant heat ready to hand, in the poultry yard at Charenton. "Until now," he noted, "chickens have hardly hatched out anything but eggs, but it seemed to me that one could just as well make them set on chrysalises and nymphs; instead of chicks, one could have the hens hatch out butterflies, flies, beetles, and so on."[39] In June, when the hens were setting, he measured the temperature in one of the nests, leaving the thermometer under the hen for seven or eight hours at a time. He found that it varied by a degree or so, depending on the stage of incubation, reaching its highest point when the chicks were about to hatch. In a rare burst of speculation, he attributed this effect to maternal emotion: "The natural heat of the hen seems to be enlivened and increased by the sentiment of joy induced by the cries of the chicks ready to hatch."[40] To tap this heat for hatching out butterflies instead of chicks, he devised a way to introduce the chrysalises into hollow glass balls, about the size of a chicken egg, with a hole in one end:

> The [chrysalises] . . . had attached themselves by the tail end to the paper cover of the jar where the caterpillars had been fed. Without detaching the chrysalis, I cut a small square of paper around the point where it was attached. I cut eight little squares of paper in this way; . . . I spread glue on the other side of the paper and I carefully applied this . . . to the interior surface of the [glass] egg. I put the chrysalises as close to each other as possible. . . . They hung down as from a vaulted ceiling. After having thus attached eight chrysalises inside the egg, I closed the opening with a cork.[41]

Cutting a notch in the cork so that the insects would not suffocate, he put the glass egg under a setting hen and kept some of the same batch of chrysalises in the original jars as a control. The hen moved the glass egg to the edge of her nest, but she kept it covered and warm, accelerating the insect's transformation from fourteen to four days. On the fourth day, Réaumur found "the first butterfly ever to be born under a chicken, and the first of this kind of butterfly to have spent such a short time as a chrysalis."[42]

Having speeded up the insect's life cycle by raising its temperature, he experimented with slowing it down by moving chrysalises to his cellar, where the air measured 8.5 degrees on his thermometer. For months, he kept them in the underground chill, checking them regularly for signs of life. A year later, some were still alive, though others had suffered from the high humidity of the cellar. During the course of the year, he kept track of different sets of chrysalises of various species in different temperatures—in the greenhouse (in winter), in the henhouse at Charenton (in June), in the workroom, and in the cellar of the Hôtel d'Uzès—to compare the time to transformation for each species in each environment. "Anyone fascinated by insects (*curieux en insectes*)," he remarked, "if he has a hothouse at his disposal, would find it charming to cause metamorphoses to happen at the beginning of winter."[43] Not surprisingly, Bazin was "ecstatic" at the prospect, and hatched all his available chrysalises in this way in Strasbourg as soon as he heard about the idea.[44]

ARTISTS AT WORK

These examples are only a few of the legions of experiments and observations pursued in the various spaces in and around Réaumur's town and country houses, in some cases extending out to other places like the royal greenhouses or the forests on the outskirts of the city. Réaumur had the essential assistance of Pitot and Nollet and other shorter-term helpers like Baron (who lived in the house) and Perreau. He also relied on several different artists to document his observations and illustrate the published text. Ideally, he said, a naturalist would have the necessary skill to make his own drawings: "The advantage of such a talent, above all, is to have the ability to capture those unique moments that leave no time for recourse to the hand of an outsider whom one cannot have continually at one's side." But Réaumur had no such skill.[45] For years, he employed two artists who also filled commissions for the academy and the botanical garden, Claude Aubriet and Philippe Simonneau. The latter, especially, made a great many drawings "under the eyes" of Réaumur, and he engraved the plates for the first three insect volumes. Simonneau had many other demands on his time, however, and could not be counted on for the kind of dedication required for the insect project. So in 1730 Réaumur decided to train a novice artist. He invited Monsieur Regnaudin, a promising young man with "a strong inclination to copy nature faithfully," to join the household as an apprentice illustrator. Little is known of this artist, though his name suggests that he may have been a relative. Installed in the Hôtel d'Uzès, he was instructed in the art of drawing, perhaps by one of the academy's artists.[46] The young man learned to use a magnifying lens and a microscope and was soon put to work drawing instruments as well as insects; the plate illustrating Réaumur's first paper on thermometers was engraved from his work (see fig 3.2). In 1731 the Academy paid Regnaudin for some thirty sheets of drawings, but a year later his untimely death cut short his career and disrupted Réaumur's long-term plan.[47] In the event, though he did use some images from Regnaudin's sheets, Simonneau engraved the lion's share of the plates for the first volume of the insect book from his own drawings, as well as a few by Aubriet.[48]

Following insects through their life cycles, and making them accessible to the eye and pencil of an illustrator, could be the most prosaic of activities. Consider this account, from Réaumur's discussion of a certain caterpillar inhabiting grains of wheat, of a minor discovery made in one of the workrooms in the Hôtel d'Uzès:

> While M.lle *** was drawing the cross-sections of these grains, to expose the silk partitions dividing into two cavities the grains whose interiors had been eaten, she needed to cut open some of these grains to better see and see again [*voir et revoir*] their partitions. She selected some that had been pierced [at one end]. She did not want to risk unnecessarily cutting the body of a caterpillar in two, and I had told her, as a certainty, that the grains with open holes no longer had caterpillars, chrysalises, or butterflies inside them. As luck would have it, in cutting into a grain with such a hole, she cut a caterpillar [inside the grain], and when this occurred for a second time, she showed me what had happened.[49]

While attempting to see minute details in order better to represent them with her pencil, the anonymous artist unexpectedly discovered live caterpillars inside the hollow grains, contrary to what she had just been told "as a certainty." This anecdote seems designed to show the artist doing considerably more than drawing what had been put in front of her. Admitting his mistake in assuming that the pierced grains would be empty, the naturalist says explicitly that the lady's persistence (she tried the experiment twice before showing the results to him) pushed him to investigate further. Perhaps, he surmised, the caterpillar had chewed the hole with its teeth, to provide a future escape route for the butterfly; sure enough, when he turned his magnifying lens on a pile of intact grains, he found many of them with a barely visible circle marking a

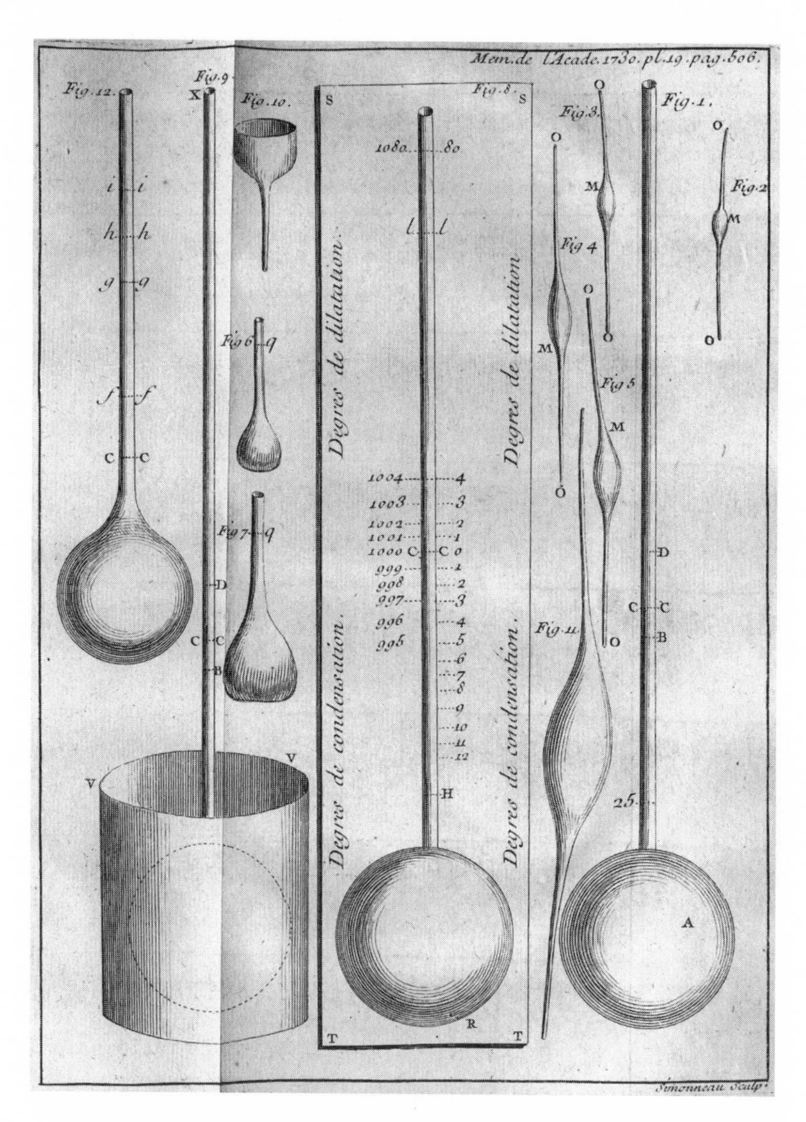

Figure 3.2. Thermometers and other glassware used in thermometric experiments. Engraved by Philippe Simonneau from drawing by Regnaudin. Réaumur, "Règles pour construire des thermomètres dont les degrés soient comparables," *Mémoires de l'Académie royale des sciences* (1730), plate 19.

hole chewed almost all the way through from the inside. The grains with open holes, with caterpillars still inside, turned out to be the rare cases where the perforation had broken through prematurely, before the metamorphosis. Réaumur's narrative attributes to the artist not only the images of the grains sliced open to reveal the caterpillars' work, but also essential elements of a novel discovery about the behavior of this insect. Though the author reserved for himself the verification of his conjecture ("I observed with a loupe"), the text tacitly reflects the back and forth of the artist and the naturalist, exchanging views, showing each other what they found, and ultimately solving the mystery of how the new butterfly could break through the hard shell of the grain.

The woman absorbed in the miniature spectacle afforded by the wheat-eating caterpillars made hundreds of sheets of drawings for Réaumur. Unlike many of his interlocutors and collaborators, her name never appeared in print, at her own insistence. He honored her request for anonymity, but she appears in the text from time to time, as in the passage quoted above, under the elliptical moniker M.lle ***.[50] These occasional mentions, along with many other traces of her activities in unpublished notes and letters, provide some clues about how the artist and naturalist worked together in Réaumur's home. His first reference to her in print comes in the preface to his first volume, published in late 1734, where he mentioned her in his reflections on the difficulty of finding the right artist to do the kind of illustrations he needed. Regnaudin had lived as part of the household, where he was being groomed for this task at the time of his death; Simonneau, though highly skilled and experienced as a scientific illustrator, could not be present whenever he was needed. In addition to drawings by these artists, Réaumur reported that his book contained a few illustrations made by a different hand:

> These are by someone of the same sex as the person to whom we owe the drawings of the insects of Surinam [i.e., Maria Sybilla Merian], someone who until now only amused herself rarely with such works, and who is so far from wishing to take some glory from them, that she does not allow me to name her. Genius and the fortunate natural aptitude for drawing, allowed her quickly to acquire a facility for making such lifelike portraits of insects that she could not help but find this work agreeable. The pleasure she took in them allows me to promise for the subsequent volumes a good number of her drawings, as faithful and as correct as one could wish.[51]

On the verge of identifying the female artist, the author drew back from naming her because she had forbidden him to do so, using polite circumlocutions to indicate not only her sex, but her gentility, her discretion, and her talent. The comparison to Merian is certainly flattering; Réaumur had discussed her work just a few pages earlier.

M.lle ***'s modest insistence on anonymity was a typical feminine stance and hints at a genteel family origin.[52] The carefully framed praise for her accomplishments graciously leaves her the choice of continuing to work on future volumes, since she draws for pleasure rather than strict necessity. The implied contrast with Simonneau suggests that she is anything but an artisan. She finds the work "agreeable" because she is good at it; this linking of work and pleasure is appropriate for a gentlewoman. But she is not exactly independent either, in spite of her natural talent. Réaumur claims credit for harnessing her talent and "genius" to the work, albeit pleasant work, of illustration. In the final sentence quoted above, he suggests that the artist herself, like her future drawings, will prove to be "as faithful and as correct as one could wish." And in fact, for all the subsequent volumes on insects, and for many other projects that never reached publication, this woman, always unnamed in print, documented and illustrated Réaumur's work, producing literally thousands of drawings. While she herself remains almost invisible, hidden behind the asterisks that replace her name, hundreds of her drawings and the plates engraved from them survive as testimony to her years of focused effort in the service of Réaumur's project. (His six published volumes include more than 250 plates, each incorporating many individual drawings; a great many other drawings were never published.)[53]

In 1735, only a few months after the publication of his first volume on insects, Réaumur wrote his last will and testament. He named as his heir Mademoiselle Hélène Dumoustier, also known as "Mlle. du Moutier de Marsigli," leaving her "all that the laws and customs permit me to give her" (which would exclude his inherited property). Apart from small legacies to a few friends and servants, he named no other heirs.[54] The testament went on to explain why he was leaving her this legacy:

> I would like to be able to express to her all the gratitude that I owe her for the use she kindly made for me, with so much patience and steadfastness, of her talent for drawing. This is what made it possible for me to publish my *Memoirs on the History of Insects* and to continue this work [*travail*]. Whatever inclination I might have had for this work [*ouvrage*], I would have despaired of finishing it and I would have abandoned it on account of the time I would have lost had I been obliged to continue to have ordinary illustrators work under my eyes. The attempts I made to do this, with a great deal of trouble, taught me that this loss of time would have been greater than one could imagine.[55]

Réaumur is adamant here, in the relative privacy of his last will and testament, about the essential contribution of the talented Mlle. Dumoustier—the partially effaced Mlle *** encountered above. He even says, rather histrionically, that he would have abandoned the massive project without her assistance. When he wrote his will in 1735, Réaumur was finishing up the second insect volume, with virtually all of its plates engraved from Dumoustier's drawings.[56]

In print, when he discreetly introduced his new artist to his readers, Réaumur commented on the usual limitations of scientific illustrators, who could not be expected to know exactly what to depict, or to see as a naturalist would see:

> Those who, like myself, are incapable of making for themselves the drawings they need, ought at least to have them made under their own eyes, whatever time it may cost. An illustrator [*dessinateur*] may be intelligent, but it is impossible for him to enter into the views of an author, if the author does not guide, so to speak, his brush. The artist will be struck by certain parts of an object that he will try to make more visible, and these will be the least important to show. It is up to the author to give the positions and the points of view of the object.[57]

This is the kind of condescension to be expected of an aristocratic naturalist toward an artist working under his instruction.

In his will, where he could refer to her by name, Réaumur praised Dumoustier's mode of work in terms exactly parallel to language in the published preface, written only a few months earlier. Exceptional not only in her skill at portraying details accurately, she also seemed to know what he wanted from a drawing:

> Mlle. Dumoustier's taste and intelligence equaling her talents, I could rely almost entirely on her. The drawings she made as I watched [*sous mes yeux*] were no more correct than those she drew in my absence. Not only did she know how to enter into my views, she knew and she knows how to predict them, since she knows how to recognize what is most remarkable about an insect, and in what position it should be represented.[58]

Thus she was not only reliable and accurate; she also had learned to anticipate what the naturalist would want depicted, and even to spot unusual phenomena for herself. This was high praise indeed from someone who had complained about the inability of artists, even the young man he trained himself, to "enter into the views of an author."

Hélène Dumoustier has always been something of a cipher in the literature about Réaumur. She has been misidentified as the daughter of the Italian naturalist Count Marsigli and as the granddaughter of Fontenelle. Jean Torlais, Réaumur's admiring biographer, was more circumspect, calling her "the collaborator, or . . . the secretary of Réaumurs," but he did not find any documents to provide any biographical details.[59] Réaumur's will is a rich source, and a remarkably explicit testimonial to the value he placed on Dumoustier's work. It says nothing, however, about how they came to work together, nor about her life and family background. Her public invisibility could easily be interpreted as evidence of her subordinate status, or perhaps of the deliberate erasing of female agency by an overbearing male author. Was she a younger woman co-opted by a prosperous unmarried academician? Did she become his mistress? Was she an artist whose ambition was thwarted by her circumstances? Who, after all, was Hélène Dumoustier?

It seems that no letters or papers in Dumoustier's own hand survive, apart from her signature on legal documents and her own last will and testament, written in the 1760s in the trembling script of an old woman.[60] Unlike the richly documented life of Réaumur, Dumoustier's biography must be pieced together from a sparse record buried in notarial archives, and even then mysteries remain. Some nuggets of information turn out to be false or misleading, even when they came from the lady herself. During legal proceedings after Réaumur's death, for example, when asked her age, she underreported by eighteen years, which could hardly have been credible to the notary taking it down; a few years later, in a different court document, she would give only an estimate of her age.[61] Baptismal records clarify her date of birth and parentage: she was born in Paris in 1691, making her eight years younger than Réaumur, and hardly a young woman when she began drawing insects for him in the 1730s. She was the fourth of five daughters born to Nicolas Dumoustier, "merchant, *bourgeois de Paris*," and his wife Hélène Potel, who lived in the market neighborhood of Les Halles.[62] Nothing in the record explains the extra name "de Marsilly," which was not used by anyone else in her family. Nevertheless, her acquaintances called her "one or the other name," according to Réaumur.[63]

Nicolas Dumoustier's commercial business cannot be traced, so it is impossible to assess the family's prosperity with any accuracy. None of the five sisters married; the two eldest took holy vows as young women and lived out their days in the Abbey of Malnoue. Their parents set them up with a lifetime income to be paid to their convent, where they lived in relative comfort as choir nuns. (Choir nuns generally came from elite families, paying their own living expenses and engaging in genteel occupations rather than manual labor.)[64] In 1721 the parents invested 1200 *livres* in an annuity for the benefit of the mother and the three younger daughters.[65] They bought the relatively modest annuity at a time of extreme financial crisis, in the immediate aftermath of the collapse of John Law's system and the Mississippi bubble; some people made enormous sums from this debacle, while others lost dramatically. Sometime in the next few years, Nicolas Dumoustier died, leaving his widow and unmarried daughters with limited resources. It is not clear how the financial turmoil of this period affected the family. At the least, it seems likely that the death of the father contributed to a constriction in the family's fortunes. This suggestive comment from Réaumur's will may refer to the mismatch between Mlle. Dumoustier's financial resources and her social status: "Everyone who knows her knows how admirable she is for the sweetness and the goodness of her manners and for the qualities of mind and heart, and that she deserved to be better treated by fortune than she was." The legacy should probably be read in part as an attempt to rectify the situation, though as it turned out, Réaumur lived for twenty-two years after making out the will in her favor; by the time of his death, she was sixty-six years old.

Dumoustier was with Réaumur when he died at one of his provincial properties in 1757. Soon thereafter, the holograph will, which he had apparently not told her about, was found in a locked cupboard in his Paris home; various distant relations then contested its legitimacy, leaving a paper trail in the judicial archives. These documents, especially the transcript of a direct interrogation of Dumoustier by a commissioner of the court, shed some light on the unspoken story behind Réaumur's testament.[66] A list of sixty questions had been drawn up by Françoise Darras, an elderly widowed cousin of the deceased, who claimed her rights to half of Réaumur's property. The widow's strategy seems to have been to impugn the integrity of the legatee. To this end, Dumoustier was asked a number of leading questions about the origins of her connection to the naturalist, the terms under which she had worked for him, and the circumstances in which she lived. Her answers, recorded by a secretary of the court in the third person, were laconic and vague about some details. Her memory for dates was not good, and her hand was visibly shaky when she signed to certify the document. Nevertheless, this interrogation provides invaluable evidence about the living arrangements of Dumoustier and her family.

Under questioning, Dumoustier explained to the lawyer what her adversary surely knew, that the Dumoustier family had originally become acquainted with Réaumur through his close friend and associate Pierre Jarosson, who was the nephew, by marriage, of Nicolas Dumoustier.[67] Hélène Dumoustier's paternal aunt was married to Jarosson's maternal uncle; she was thus, in the parlance of the times, his aunt *à la mode de Bretagne*, or his second cousin by marriage.[68] Jarosson lived and worked with Réaumur in Paris in the 1720s, as we have seen, and their families in Poitou were closely acquainted. Jarosson must have introduced him to the Dumoustier family early in the 1720s; scattered evidence suggests that the women were frequent guests of Réaumur after the death of Nicolas Dumoustier. In a letter written in 1723 from Cosne, where he was supervising the iron foundry, Jarosson sent his regards to "your Charenton ladies," who can plausibly be identified as the Dumoustier women.[69] Just a few weeks earlier, in October 1723, the Réaumur parish register recorded the baptism of a baby whose godparents were none other than Hélène Dumoustier and René-Antoine Ferchault de Réaumur. Quite a number of witnesses signed the register, including Jarosson's mother, Jeanne Elisabeth Thibault (Hélène Dumoustier's aunt), and several of Réaumur's relatives.[70] Thus the families were evidently on cordial if not intimate terms well before she began drawing insects.

When questioned in 1759, Dumoustier recalled that she and her family had moved into the Hôtel d'Uzès in the early 1730s. The notarial archives corroborate her recollection: in 1735 her mother leased the "little Hôtel d'Uzès" from Réaumur.[71] They may have stayed there earlier, but the lease formalized the arrangement, with the women paying rent for a building across the courtyard from the main house. The two-year lease was drawn up within a few days of the date of Réaumur's will, while the second volume on insects was in preparation, and would have ended when the lease on the larger house expired in 1737. As it turned out, the Count d'Uzès was forced to sell the property in 1736 to pay off his debts, and the occupants had to move precipitately to a house in rue neuve Saint Paul, near the Arsenal.[72] Just a few years later, in 1740, they all moved again, with the laboratory equipment and the expanding natural history collections, to a house on rue de la Roquette in the Faubourg Saint-Antoine, just outside the old city walls.[73] Throughout this period, the women accompanied Réaumur on his annual visits to Poitou, as they became part of his intimate social circle. They were on friendly terms with other frequent visitors like Bazin, who reported drinking a toast with neighbors "to your health and to that of Madame and Mesdemoiselles Dumoustier" after their departure from Poitou in 1733.[74] Dr. Baron's wife, also in Poitou, asked Hélène Dumoustier to buy her some fashionable lace only available in Parisian shops.[75] In addition to frequent stays at the country house in Charenton, the party sometimes traveled to Malnoue, a few hours journey by carriage, where the older sisters lived as nuns.[76]

When asked years later about the capacity in which she had lived in Réaumur's house, Dumoustier replied, "as a friend occupying an apartment rented by her mother." After the death of her mother in 1743, she and her sisters had continued to live in their apartment "as friends and companions keeping their home [*ménage*] distinct and separate . . . at their own expense, just as their mother had done."[77] Under further questioning, she reiterated that she and her family had maintained their living quarters at their own expense; they had a small kitchen and their own servants. Dumoustier always emphasized her family's separate space and financial independence.[78] In another deposition taken during the inventory process, when challenged about her right to stay in the house after Réaumur's death, she was asked whether she had paid rent. According to the notary:

> She said that during the time she had lived in the house . . . where her mother and one of her sisters had died, she had not paid any rent, that she was not "en pension" in the deceased's home, that she maintained her household and that she ate at Mr. Réaumur's only when he was eating there and asked her to join him, and that the furniture in the space she and her sister occupied in the said house belonged to them, and she offered to show us the apartment and her kitchen.[79]

Jarosson also had a room in the house on rue de la Roquette until he married in 1742. So for a time there were effectively three linked households under one roof in Faubourg Saint-Antoine. "Seeing that there were three masters lodging together in the same house—the said Réaumur; Jarosson, *secrétaire du roi*, nephew of the mother of the respondant; and the respondant and her family—and since the respondant and her family could not have an apartment as large as those they had occupied previously . . . she said Réaumur did not think it appropriate to make them pay rent for it."[80]

Some parts of the 1759 interrogation seem intended to force Dumoustier to admit her poverty and dependency. Questions about how she learned to draw, and for what purpose, implied something shameful about her work as an illustrator and a lower status than she would accept for herself. "Is it not true," one of these questions asked, "that as she had no patrimony, in giving herself to this talent, she counted on this same talent to be useful to her and to supply her needs?" To this she replied, with some hauteur, that "she is not obliged to account for her fortune, that it suffices to say that her family left her with the means to live without being obliged to work for her living, and that if she did learn to draw, it was to satisfy her taste and to be an amusement for her." She said she had studied "miniature" drawing for eight months "in her tender youth," simply to please herself, never imagining using her skill to earn money. She abandoned drawing "out of youthful laziness" and only picked it up again when she saw artists (probably Regnaudin and Simonneau) at work in Réaumur's house; she had no formal instruction apart from that one drawing master when she was a girl. Of course, given the adversarial nature of the proceedings, this may not be the full story. She may well have been a frequent visitor at the time when Regnaudin moved in, and she could even have been present for his drawing lessons. She certainly learned how to use the loupe and microscope for magnified images as she became familiar with natural history and became adept at depicting insects in their various guises. By 1759 she probably had only a hazy recollection of how she had come to spend so much time drawing.

In any case, as she answered questions from the court, Dumoustier stubbornly resisted the insinuation that she had taken up drawing as a source of much-needed funds. Refusing the suggestion that she had somehow turned her connection to Réaumur to her own financial advantage, she insisted that she had acted throughout as a lady of leisure, a genteel companion and not a dependent client. Asked pointedly "whether Réaumur had had the Academy pay for her drawings, or paid for them with his own funds," she recalled one instance, "while her mother was still alive," when Réaumur had given her the sum of fifteen hundred *livres* from "the king's funds" for her drawings of insects. She even recalled her surprise, "because she did not imagine that her drawings should produce anything for her, having made them only for her own amusement and to please the said *sieur* de Réaumur; she invested this sum at the time in annuities, and declared that it was the only sum she received from said *sieur* de Réaumur."[81] Since she did not expect monetary reward or payment for her drawings, she implied, she did not belong in the world of artists who lived off their talents.

There is no reason to doubt Dumoustier's representation of herself and her family as close friends of Réaumur living in their own rooms in his various residences and participating in the social life of the house. By the time of his death, Hélène and her sister Magdeleine had been part of the extended household for over twenty years.[82] It is also clear, however, that she either misremembered or misrepresented at least some of their financial arrangements. Records in the archives of the Academy of Sciences show unequivocally that she was paid for hundreds of drawings starting in 1736. Generally, the artwork for academy publications was documented with detailed memoranda of drawings and engravings, paid at a standard price per sheet. Each sheet included multiple finished drawings, depending on the subject; the individual drawings selected for engraving would later be cut out, rearranged, and glued onto the sheets from which the engraver worked. Normally the artist's name appeared

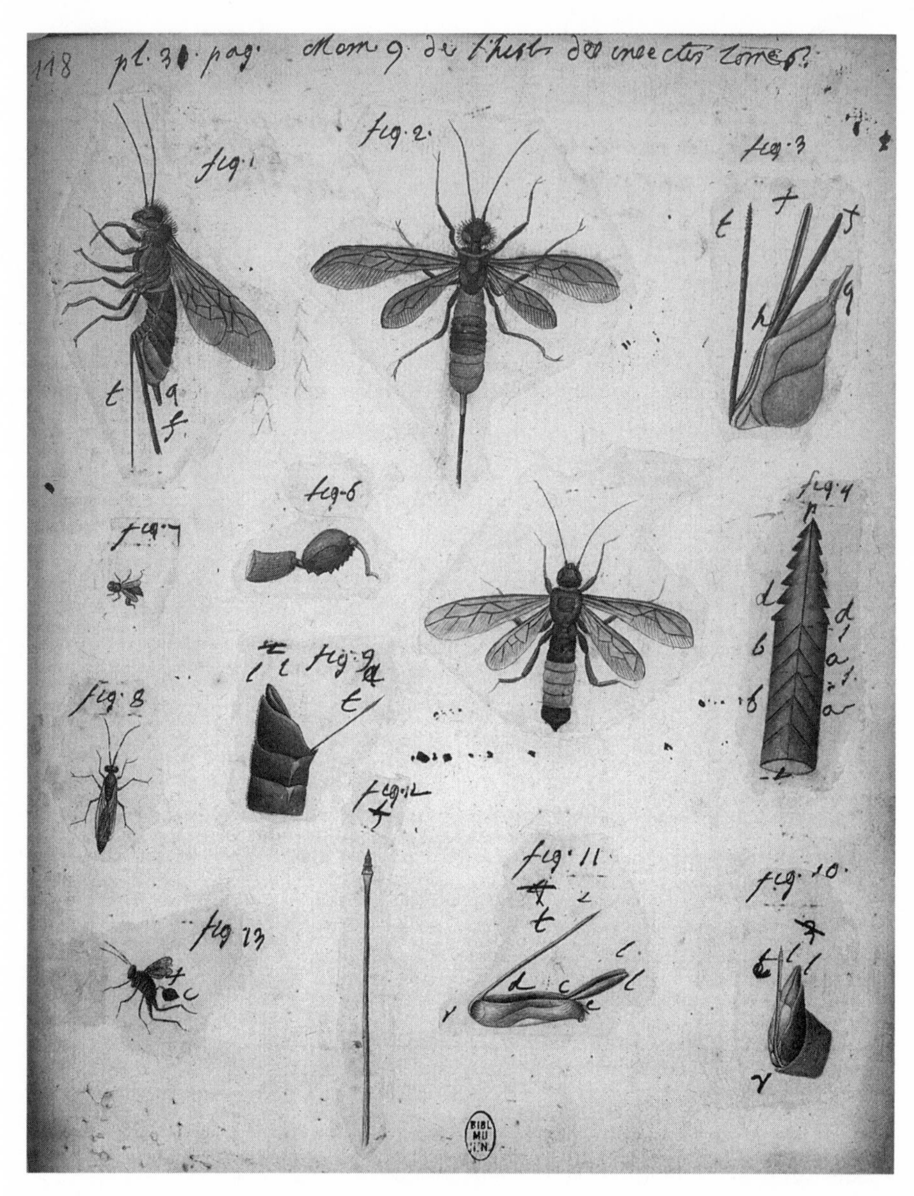

Figure 3.3. Typical sheet of drawings prepared for engraver. Drawings by Hélène Dumoustier cut out, arranged, and glued onto page; figure numbers inserted by Réaumur. Multiple views of several different kinds of stinging flies, including "a large female ichneumon fly brought from Lapland by M. de Maupertuis" (figs. 1–5) and another from Île de France (fig. 13). Published as plate 31, Réaumur, *Mémoires pour servir à l'histoire des insectes*, vol. 6. Ms. 1901, BCMHN. By permission of Bibliothèque centrale du Muséum nationale d'histoire naturelle, Paris.

on these memoranda of drawings and the academy's treasurer paid the artist directly. Dumoustier's drawings are itemized in the same way as the others, but she was not identified by name; the cost was reimbursed to Réaumur rather than paid directly to her. Over the period from 1736 to 1747, receipts survive for 350 sheets of these drawings, described in sufficient detail to attribute them unequivocally to Dumoustier. She was paid at the same rate as Simonneau, 25 *livres* per sheet.[83] Thus she received at least 8,750 *livres* over eleven years, far more than the one-time payment she recalled for her inquisitors in 1759. What are we to make of this oversight, or possibly outright prevarication? The archives yield no definitive answer, but it seems likely that Dumoustier was protecting her genteel social status, whether consciously or not, from any suggestion that her straitened circumstances had driven her to work for money, like an artisan. In the same vein, her family connection to the wealthy and socially mobile Jarosson would also have lent credibility to her claim to gentility.

HÉLÈNE DUMOUSTIER'S OBSERVATIONS

The uncensored testimonial to Dumoustier's talent and hard work in Réaumur's will went beyond his appreciation of her drawings and her artistic skill. He also acknowledged her contributions to the steady stream of observations and experiments that filled their lives: "I ought as well to do her a justice, which her modesty did not permit me to do in my published works, however much I owed it [to her] and would have wished to do so: she supplied me with several observations that became part of my books and that [otherwise] would have escaped me."[84] Tracking the elliptical references to these observations in print and in manuscript notes brings to light Dumoustier's intimate familiarity with the minutiae of natural history and puts her into the scene wherever observations were being made in and around the house. Réaumur's discursive strategy in print—to show observations being made, rather than simply reporting results—affords occasional glimpses of the artist behaving as a naturalist, as in this attribution: "This gall was first observed by Mlle du ***."[85] They were taking a walk, looking doggedly for insects inside the galls on oak trees:

> I opened several without finding anything, and M.[lle] *** discovered a worm in one of the first that she opened; subsequently she and I both found some in almost all the galls of that kind that we examined. . . . We often saw one of them and were uncertain whether it was a worm until it moved.[86]

The first-person plural pronoun, used very rarely in print, gives this passage a decidedly personal tone, pointing to the companionable nature of the outing.

Figure 3.4. Hélène Dumoustier, galls on oak leaves. Drawings selected and prepared for engraving. Published as plate 35, Réaumur, *Mémoires pour servir à l'histoire des insectes*, vol. 3. Ms. 1901, BCMHN. By permission of Bibliothèque centrale du Muséum nationale d'histoire naturelle, Paris.

✵

Sometimes these moments happen while Dumoustier is drawing; at other times, she becomes just another collaborator or informant witnessing this or that detail:

> These flies have a quality that I have not seen in others of their kind . . .: they have an odor that appeals to cats. While Mlle *** was drawing one of them, with others of the same kind nearby enclosed in a paper, a cat came up onto the table, and immediately started to rub the end of its nose, and alternately one and the other side of its head, against the paper where the flies were enclosed.[87]

If she had not been working in a room with a cat, they might never have noticed the smell of the insect. This sort of passing reference anchors her drawings in her continual engagement in the scientific life of the household. Dumoustier was nearly always there, ready to turn her pencil or pen to whatever task presented itself. Years later, when asked why she went along to Poitou every year, she insisted that she was no frivolous companion, that she was "even necessary to him because of her drawings, for which he was always finding objects."[88]

In the case of the pine caterpillar discussed in chapter 2, Réaumur made a point of attributing to Dumoustier the observation of an odd structure, discovered while making her drawing of the hairs growing from the rings of the caterpillar's body. "This detail, well worth remarking upon, had escaped me when I first had some of these caterpillars."[89] While examining the arrangement and color of the bristles growing in little tufts out of each ring, she noticed a tiny oval cavity, bounded by a slightly raised cordlike structure, on the top of each segment of the body. To her surprise, the shallow cavity opened and closed as the caterpillar moved:

> M.lle du** observed in it another peculiarity: the inside [of the cavity] was filled with a cottony material, formed of short hairs. While the caterpillar moved, and while it opened and closed this vent [*stigmate*], little puffs of this cotton rose above the edges of the cavity; they seemed to be no longer attached to the body . . . and sometimes they were even shot up to some height.[90]

This description must have come from the artist herself, since Réaumur was not present when the puffs of white fibers were first observed. Dumoustier was working with live caterpillars; he specifies that he had given her a batch of caterpillars recently emerged from the nest and left her to observe them by herself. Having been alerted to the peculiar phenomenon, Réaumur made a point of looking for the oval cavities when a new nest of caterpillars arrived from Bordeaux: "I was attentive to observe them, and I saw the play of puffs of cottony hairs." A few days later, the cavities were no longer filled with these little hairs, and he could see the surface beneath.[91] This ephemeral detail—only readily visible when the young caterpillars first emerged from the nest—would have been missed if not for the artist's sharp eyes. In order to draw its bristles, she had to examine the caterpillar in ways Réaumur had not done, and she saw something unexpected. A trivial matter, perhaps, but for our purposes an interesting one, as Dumoustier appears repeatedly in these few pages, and the text hints at the dynamic way the naturalist's and the artist's observations play off each other, leading them to investigate further the mechanism that sent the

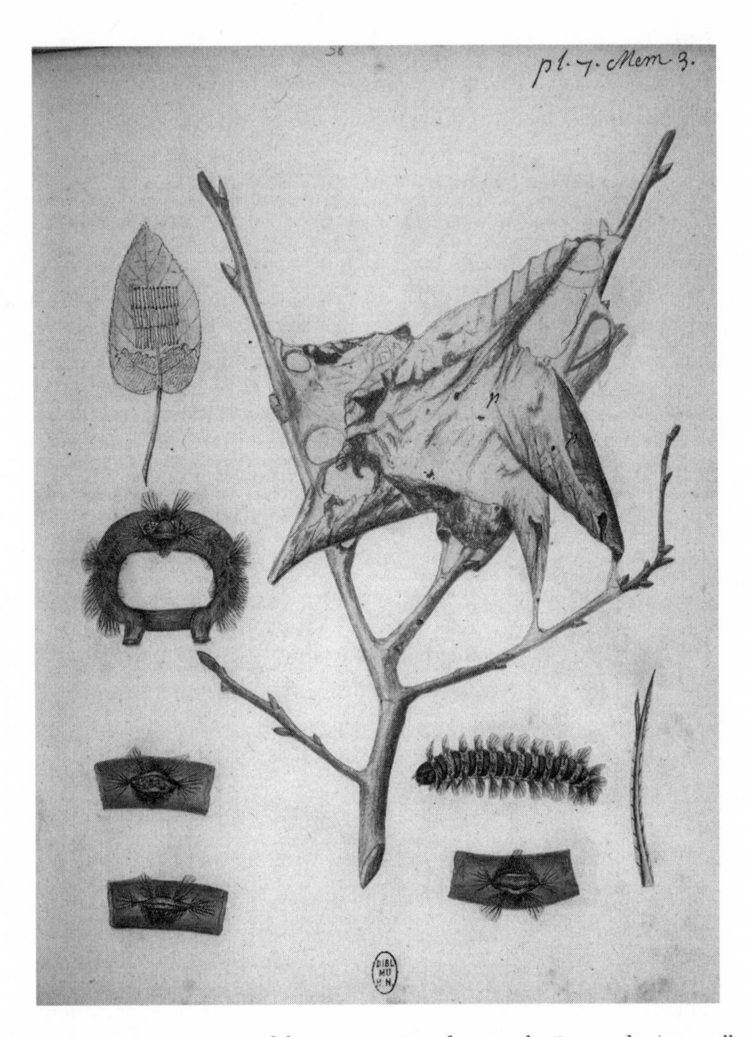

Figure 3.5. Drawings prepared for engraver. Two drawings by Regnaudin (caterpillar nest of folded leaves, leaf with newly hatched caterpillars); all views of pine caterpillars by Hélène Dumoustier. Enlarged details show vents and bristles; the two images in lower left show the vent open and closed, with puffs of fiber just visible in the opening. Published as plate 7, Réaumur, *Mémoires pour servir à l'histoire des insectes*, vol. 2. Ms. 1901, BCMHN. By permission of Bibliothèque centrale du Muséum nationale d'histoire naturelle, Paris.

little hairs into the air. Far from discounting what she saw when he was not there to confirm it, Réaumur assumed that he would be able to find the novel phenomenon in other caterpillars. Clearly, he trusted her to make the most delicate observations.

In the spring of 1736, mating frogs in a pond on the grounds at Charenton captured the attention of the household, and the creatures were studied there every spring for the next five years.[92] Although these observations were never published, the surviving manuscript notes and drawings give us yet another vantage point on the integration of this kind of investigation into the daily routines at the country house. The questions or conundrums driving the investigation were remarkably like those asked over and over about insects: mating behavior and the mechanism of fertilization and egg-laying. Frogs, because of their size and the long immobility of mating pairs, were easier to see and to watch than insects. But the peculiar features of their coupling—especially the oddly textured gland on the male's hand where it gripped the female's torso, the lack of an obvious male sex organ, and the submersion of the hind quarters of the frogs in water—kept the whole process remarkably opaque. To make matters worse, Swammerdam had also observed mating frogs, more than fifty years earlier, and had depicted the male spraying the egg mass with seminal fluid. Try as he might, Réaumur could not see what Swammerdam had seen, leaving him to wonder if the nuptial pad on the frog's hand might not be somehow essential to fertilization.[93]

Years later, in a letter to Lazzaro Spallanzani, Nollet recalled his participation in the studies of the Charenton frogs.

> M. de Réaumur and I did considerable research on this matter. We followed, with a great deal of care and patience, those embraces and copulations for weeks; I remember putting little pants [*caleçons*] of waxed taffeta on these small animals, and watching them for quite some time, without ever being able to see anything that indicated an act of fertilization.[94]

These garments occupied them for most of one mating season as they designed one experiment after another to determine the source of insemination. The first pants were made of animal bladder, stretchy enough to slip easily onto the frog, but ultimately unsuitable for its tendency to shred and soften in water. The waxed taffeta, a waterproof material normally used for such things as umbrellas, held up better, but was less flexible and more difficult to keep on the animals. On the first try, the frogs pulled their legs up inside the pants, pushed off, and jumped right out of them, as the frustrated naturalists watched. The

Figure 3.6. Hélène Dumoustier, mating frogs. The male is wearing waxed taffeta pants, with straps passing over its shoulders. Ms. 972, BCMHN. By permission of Bibliothèque centrale du Muséum nationale d'histoire naturelle, Paris.

❋

problem was eventually solved by making the leg holes smaller and taking a few stitches to adjust the fit after the frog was encased in the garment. "But what ensured the whole thing was that I put straps on these pants. I passed them over the arms of the frog, under the head, and between his body and that of the female." Dumoustier was present for these maneuvers and documented the pants with their little straps with her pencil and brush.

Though Spallanzani later succeeded in retrieving seminal fluid from inside similar garments, and even managed to artificially inseminate frog eggs with it, Réaumur and Nollet never got definitive results of any kind.[95] After the initial season, Nollet was replaced by other assistants. The idea of clothing the frogs was dropped, as was speculation about the function of the nuptial pads. By 1740, the last season in the Charenton house, no one had yet witnessed either the emergence of the eggs or anything indicating how fertilization was accomplished, even though a whole collection of mating pairs had been kept in a container under a large bell jar on Réaumur's writing desk. "Some of them even produced their eggs while I was working next to them on my desk, without my having seen them emerge. What I have learned at least is that they lay the eggs in less than a quarter of an hour, perhaps considerably less."[96] Réaumur had dissected frogs and found a milky substance in their testicles, but had never found evidence of such a liquid in the water around egg masses. Fortuitously, his notes for this year give considerable circumstantial detail, allowing us to put several people into the scene. His new plan was to divide up twelve pairs of frogs among three observers, with one pair to a jar. Réaumur kept eight of these for himself, gave two to Dumoustier and two to Jean-Etienne Guettard, a young protégé just getting started on a long career in natural history.[97] Guettard's observations in this instance were not recorded, for some reason. Dumoustier missed the event in one jar, where the eggs appeared while she was out of the room, but her redoubled attention to the remaining pair was amply rewarded when she noticed eggs emerging from the female:

> Instantly, she did as I had recommended and turned her eyes to the hind end of the male and fixed them on it. Hardly had she focused on it when she saw a jet coming out of it, which she could only compare to what it most resembled, namely a jet of pipe smoke. As it left the male's rear end it was the size of the quill of a feather, and a bit farther out it divided into a great number of jets of finer filaments, similar to the way a plume of smoke divides itself. This only lasted an instant and that was all she could see.

It was immediately apparent to all concerned that this ephemeral phenomenon, barely visible, was what they had been looking for.

It remained only for Réaumur to see it for himself, after hearing his companion's lively description of the swirl of turbulence moving through the water over the eggs. After various misadventures, when he was down to his last pair, he got lucky. He had the jar on the chimneypiece, at eye level, with the back ends of the frogs turned to the light. "All of a sudden, I saw a mass of eggs appear; the male gave a croak, he arched his back a bit and pulled in his legs at the same time, then he stretched out and retracted his flanks. He repeated these movements three times in a row."[98] In spite of all this action, Réaumur did not see the fluid jet observed by Dumoustier and had to be satisfied with his partial observation; he noted that the eggs, kept in the jar, never developed.

Another long-term project based at Charenton was the difficult problem of the reproduction of bees and the seasonal cycle of the life of the hive. Réaumur developed techniques for paralyzing and counting and sorting bees, but he did not work by himself:

> I had with me a person who loves natural history, and who supplied me with observations recorded in the preceding volumes, and in addition to observations, very perfect drawings; a person who knows as much as I do about bees of different sex; she has made drawings of them. We set out, she and I, to examine them, and to sort them, so to speak, one by one, with more care than one gives to sorting coffee beans.[99]

They were looking for a queen, which they eventually found, and determined that she was the only one in the hive. "We were no less attentive to search for males; but in spite of all our scrupulous attentions, we could not find a single one."[100] Dumoustier's eye for detail, and her willingness to devote her attention to exacting problems, served Réaumur well in his study of bees. He also entrusted her with pursuing observations when he had to leave an experiment in progress. When they finally got bees to mate in a jar, for example, he could not stay to watch as long as he might have wished:

> After observing these proceedings, and having seen them repeated for more than two hours, I was obliged to leave my two bees and the countryside to go to Paris, where one of our academy meetings called me. But several people whom I left at my house, and one especially, whose eyes I trust as much as my own, did not cease observing what was happening for the rest of the afternoon, and upon my return they gave me an account of all they had seen.[101]

On other occasions, the two simply observed whatever crossed their path. One brilliant November day, "when the sun had shone all morning, inviting spiders to spin," they went searching for interesting spiders. In this case, the vignette comes from manuscript notes, so Réaumur does not mask his companion's name:

> As Mlle. Dumoustier and I were examining the threads going from trees to bushes, . . . while we were looking for the spiders that had stretched these

> threads [across such a distance], trying to see them performing the operation, Mlle. Dumoustier was surprised to see a spider take off from a little twig and lift up into the air as a fly would have done. She cried out that a spider was flying, and did so quickly enough that I could still see it as it flew up in the breeze.[102]

Again, the notes give just a fleeting view of the pair out chasing spiders and trying to determine how they could span such long distances with the strands of their webs.

These few examples begin to fill in some outlines of the scientific work that engaged Dumoustier for many hours each day. She knew Réaumur's collaborators and assistants well, in some cases working alongside them and forming lasting personal ties.[103] She and Bazin, for example, shared supplies and techniques; after he had moved to Strasbourg, he sent her a new block of Chinese ink to repay her for the piece she had given him a few years earlier when they were both making drawings in the manor house in Poitou.[104] She was also well known to the many people who came to the house for social visits or for tours of the collections.[105] Even in print, where he left her unnamed, Réaumur dropped enough hints to reveal her more or less constant presence when observations were being made. Her involvement can sometimes be inferred even when she is only mentioned as "a person," or as one of a group of "those who were with me."

Knowing that she was usually recording every step of most investigations with pencil and ink, we can put her into Réaumur's discovery narratives even when he does not mention her explicitly. In a study of mayflies, for example, illustrations of the insect's life stages, engraved for the published account, map onto a narrative about how they were witnessed. The house at Charenton, with its garden sloping down to a flight of steps and a landing at the river bank, was ideally situated for a study of these aquatic insects. The local species of mayfly live as nymphs, in holes burrowed into river banks, hatching out on summer evenings as swarms of delicate winged flies:

> This spectacle can hardly fail to be surprising the first time one sees it, and is more and more surprising to someone who thinks about it when he sees it again three or four days in a row. Even if one knows the origin of these flies—that they come from insects which, after having grown in the water, metamorphosed during the preceding nights—it is still difficult to conceive how so many insects could have emerged from the river at once, as if in unison.[106]

To decipher the cause of such a synchronized hatch, Réaumur got a local fisherman to take him up and down the river when the water was low, to inspect

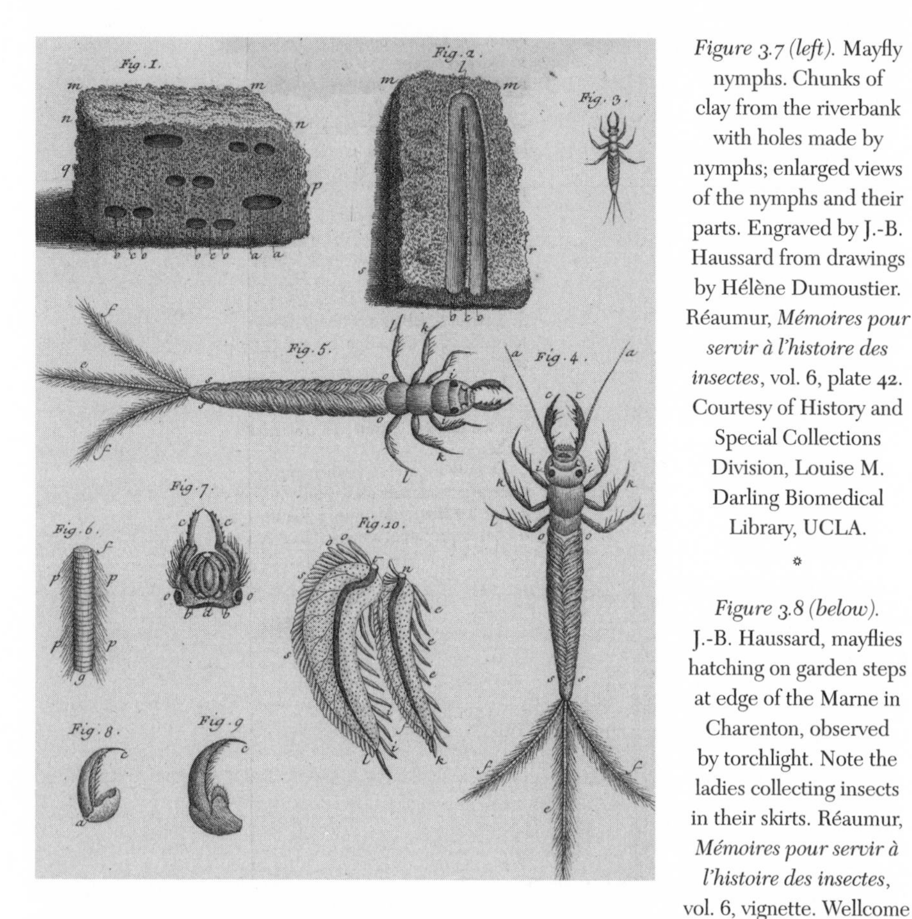

Figure 3.7 (left). Mayfly nymphs. Chunks of clay from the riverbank with holes made by nymphs; enlarged views of the nymphs and their parts. Engraved by J.-B. Haussard from drawings by Hélène Dumoustier. Réaumur, *Mémoires pour servir à l'histoire des insectes*, vol. 6, plate 42. Courtesy of History and Special Collections Division, Louise M. Darling Biomedical Library, UCLA.

Figure 3.8 (below). J.-B. Haussard, mayflies hatching on garden steps at edge of the Marne in Charenton, observed by torchlight. Note the ladies collecting insects in their skirts. Réaumur, *Mémoires pour servir à l'histoire des insectes*, vol. 6, vignette. Wellcome Library, London.

the mud banks, peppered with holes. They dug out chunks of the mud beneath the water line and found holes occupied by worms or nymphs, in great enough numbers to explain the storms of mayflies. Back at the house, Dumoustier drew the mud blocks from different angles and studied the nymphs under the microscope for detailed drawings of their gills and pincers.

In the summer of 1738 Réaumur attempted a closer observation of the transformation of the mayfly nymphs, and he invited guests to come from the city to see the spectacle. On a day chosen by the local fisherman as a likely time for the hatch, the naturalists went out on the river to collect clods of earth with the nymphs still inside. After dark, with lightning flashes in the distance promising a thunderstorm, they returned to the landing where the guests were waiting. As soon as the gardener set the tub down on the bank, the mayflies started hatching all at once. "I promptly seized one of the torches . . . and ran to the tub. I saw on the top surfaces of the clods of earth, where they were not covered by water, mayflies just starting to leave their old shells, others closer to emerging, and others that had succeeded in getting out and taking flight."[107] The tub made the individual transformations visible; a little later in the evening, the assembled guests watched the full spectacle unfold on the river itself:

> The number of mayflies filling the air above the current of the river was neither expressible nor conceivable. . . . The air around us was full of mayflies. . . . Several times I had to leave my spot and retreat to the top of the steps, no longer able to withstand this rain of mayflies, which . . . struck my face continually from all directions, in the most troublesome way; mayflies got into my eyes, my mouth, and my nose.[108]

The spectators watched as the insects swarmed around the lights. Anyone holding a light soon had his clothing covered with insects.[109]

The evening by the river was the climax of the investigation, even though the spectacle could not answer all questions about the reproduction of the mayfly. Careful observers could catch sight of the females laying their eggs, but it was more difficult to see the insects mating and to determine when the eggs were fertilized.[110] The illustrations show the fly emerging from the nymph and the egg packets emerging from the female, but not the mating of the adults. The images capture specific moments in the complex and sometimes surprising sequence of events. In order to represent these very brief events (the unfolding of the adult fly's wing, or the emergence of the eggs), Dumoustier had to freeze time to make the elusive phenomena appear clearly visible on the page. The drawings could not have been done during the spectacular moment by the riverbank, where it was dark and difficult to see detail.[111] In some cases, the

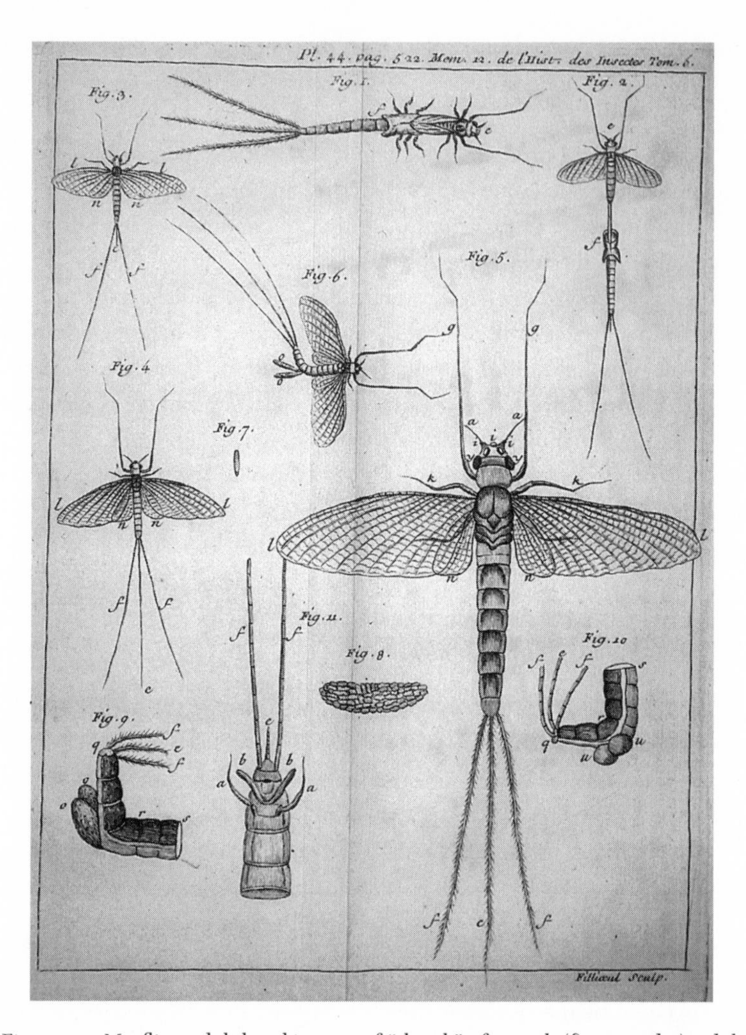

Figure 3.9. Mayflies: adult breaking out of "sheath" of nymph (figs. 1 and 2); adult male (fig. 3) and female (fig. 4); eggs emerging from female (figs. 6 and 9); egg packets, life-size and enlarged (figs. 7 and 8). Engraved by Fillioeul from drawings by Hélène Dumoustier. Réaumur, *Mémoires pour servir à l'histoire des insectes*, vol. 6, plate 44. Courtesy of History and Special Collections Division, Louise M. Darling Biomedical Library, UCLA.

artist worked from dead specimens, and sometimes the insects were manipulated to slow things down:

> I have often tried to stop the progress [of the metamorphosis] to better see how each part was housed in the case from which it was ready to emerge. I grasped a fly which had only just started to disengage its head, I pressed the head in the very moment when it had just showed itself; sometimes I pushed the cruelty so far as to flatten it and crush it between my fingers: the metamorphosis that I wanted to interrupt completed itself in spite of me.[112]

Réaumur kept some of the adult flies in jars, where they lived out their short lives; the females laid their eggs wherever they happened to be, in double clusters (see fig. 3.9). Following along, via text and image, a reader could imagine the worktable where observers manipulated the flies, put the eggs under the microscope, and speculated about the function of different anatomical features. In this long chapter on mayflies, Réaumur foregrounded his own actions with the first-person singular prounoun, but the illustrations show that Dumoustier was in the thick of the investigation for much of the time, recording fleeting events as best she could.

Most of the scholarship on making natural knowledge in the home focuses on the family. Wives and daughters assist in astronomical observations; women distill and compound medicines in their kitchens; sons might be trained to follow in their fathers' footsteps.[113] In eighteenth-century Paris, well-established scientific families like the Cassinis and the Geoffroys and the Jussieus produced generations of astronomers, chemists, and botanists. The details of domestic arrangements varied enormously, depending on available financial resources and on the kind of scientific work. Réaumur presents an interesting case because he chose a life of science as an alternative to his default role as noble paterfamilias; remaining a bachelor was in part a career decision. He was certainly attached to his ancestral home, and incorporated natural history into his sojourns there, but he never contemplated giving up his scientific life in the capital. Life as an academician took the place of marriage and the production of progeny, even though this meant the end of his family line. On the other hand, Réaumur functioned as the patriarch of an elaborate scientific household, training a cadre of assistants, artists, and collaborators, many of whom carried on his scientific legacy. Affective and professional ties grew out of the day-to-day operation of taking care of insects, operating microscopes and air pumps, building and reading thermometers, preserving and arranging specimens. For the assistants who passed through the house, these ties carried over into correspondence, collaboration, and friendship long after they no longer lived on-site. The women in the household—an all-female family embedded in this patriarchal domain—maintained their independence to some degree, but they also integrated themselves into the social and even the scientific life that filled most rooms of the various houses they occupied. Certainly the work of Hélène Dumoustier was essential to the whole operation, as Réaumur recognized.

Recruiting Observers and Training "Philosophical Eyes"

IN THE OPENING ESSAY of his book on insects, Réaumur made a special point of acknowledging Count Pajot d'Onsenbray, the director of the French postal system. Natural history depended on the benevolence of the postmaster general, who guaranteed the timely delivery of letters and packages filled with everything from observations and experimental reports to live insects and prepared specimens. Applauding d'Onsenbray's "love for the progress of the sciences," Réaumur noted that a special understanding between the two men allowed everything from simple envelopes to elaborate boxes to arrive at his house, free of charge. Réaumur described this arrangement not only to express his gratitude to d'Onsenbray, but also to inform readers—especially those he did not know personally—that they too could send insects to Paris without fearing delays or causing expense to the recipient:

> I will gratefully accept presents from all those who take an interest in the progress of natural history . . . and will gladly make my thanks public, if they will be so kind as to send me insects that they judge to be unusual. . . . They have only to enclose them in little boxes with the necessary food for the journey, and to address them to the Count d'Onsenbray, postmaster general of the French Post; inside the first envelope, they should put a second one addressed to me: they may be sure that the package will be faithfully and promptly delivered.[1]

In the same paragraph, the naturalist publicly recognized several of his most prolific correspondents, introducing some of the observers whose names would crop up in the pages to come. And he promised to do the same for future donors.

Réaumur's work on insects appeared one volume at a time over the course of eight years; another large volume was virtually complete at his death but remained unpublished until the twentieth century.[2] In part, logistical problems dictated the drawn-out production process, and the press of other work continually interfered with the writing. But Réaumur kept the project open-ended partly by design, as he explained in the first volume's preface: "I will be able to profit, for subsequent volumes, from the insights that anyone would like to communicate to me. I will be able to clarify, rectify, and correct anything that seems to require it."[3] As each volume appeared, the author acknowledged in print the contributions of his informants and collaborators. He also took the opportunity to respond to critics, to correct mistakes readers had pointed out, and occasionally even to announce discoveries made by other people. Although the nicely printed books gave the appearance of permanence, they were, like the work of observation and experiment they recorded, open to revision and correction. The volumes invited readers to look at nature for themselves, to make their own experiments and collections, and to report back to the author on anything noteworthy. Those who did report back might find their observations in print when the next volume appeared. Réaumur intended his accounts of insect life to be authoritative but not absolutely so; he invited his readers to develop a critical sensibility in assessing descriptions of the ubiquitous "singular" facts afforded by the insect world. As an ongoing (and perennially unfinished) project, the books exemplified the dynamics of knowledge production, while participating in the reciprocal exchanges that fueled so much of natural history.

Reading and writing were as integral to doing natural history as collecting and watching and dissecting insects. This may seem a trivial point, but it reminds us that books, along with periodicals (often containing lengthy extracts of recent books), letters, specimens, and drawings, circulated through a dispersed community of naturalists. We can consider the volumes of Réaumur's *Mémoires pour servir à l'histoire des insectes* from different points of view—as physical objects, distributed by the author to his correspondents and colleagues around the realm and around the world; as academic science opened up for a wider readership; and as a record of practice that exemplified how to be a naturalist. In his appeal to readers, Réaumur was extending the practice we have already seen at work in his correspondence with provincial observers eager to participate in his project and to emulate his practice. As Bernard de Fontenelle, perpetual secretary of the Academy of Sciences, noted, "The art of making these observations is quite interesting [*curieux*] in itself," thereby recognizing that one hallmark of Réaumur's manner of exposition was to reveal to readers all the circumstances, tricks, techniques, and tools that went into producing histories of insects, as well as the details of insect life.[4]

When Réaumur finally sat down to arrange his observations of insects in order to narrate their histories for publication in book form, he had been engaged with natural history for decades.[5] Starting with a study of the growth of shells in 1709, he had presented numerous academic papers on various sea creatures (torpedo fish, sea stars, sea urchins, crayfish) and on insects. Unlike the more abstruse work of the mathematicians and chemists, papers on the regeneration of crayfish claws or the sharp sensation produced by the electric organ of the torpedo fish could easily engage the attention of the audience at the academy's public sessions. He designed the books to draw people into natural history, as he had done in person at so many public sessions of the academy and for the frequent visitors to his town and country houses. Once in print the books generated even more correspondence, which fed back into subsequent volumes and the miscellaneous manuscripts that piled up in his study. The books became a standard reference on insects for several generations at least, and Réaumur himself became a paragon of observational technique and style.[6]

Corralling his mountains of observations of insects into book format, Réaumur chose the genre of the "*mémoire*," modeled on the academic paper, as the most appropriate form for his histories.[7] Each volume, like a volume of the academy's journal, was a collection of these papers, which could be read separately or in sequence. The *mémoire* was essentially a report, whether mathematical, observational, or experimental, that very often incorporated some element of first-person narration or framing. The history of a wasp or a moth, for example, would weave the story of how the insects had been observed into the story of how they lived, what they looked like, and where they could be found. "Anyone who announces [singular facts] for the first time cannot overemphasize that he saw them, and how he saw them: hardly any form but the *mémoire* allows one to speak often in this tone."[8] Réaumur highlights method here, but also "tone," a style intended to keep the observer in the narrative, to draw the reader into the scene, while enabling the reconstruction of a similar scene should the reader care to make his own observations. No detail of circumstance or technique was too trivial in the service of transparent reporting. The reader must know that the author guaranteed every detail through his own experience, by recounting what he had seen for himself. Statements based on "hearsay" would throw the credibility of the whole account into question—and the literature of natural history was replete with such hearsay or incomplete

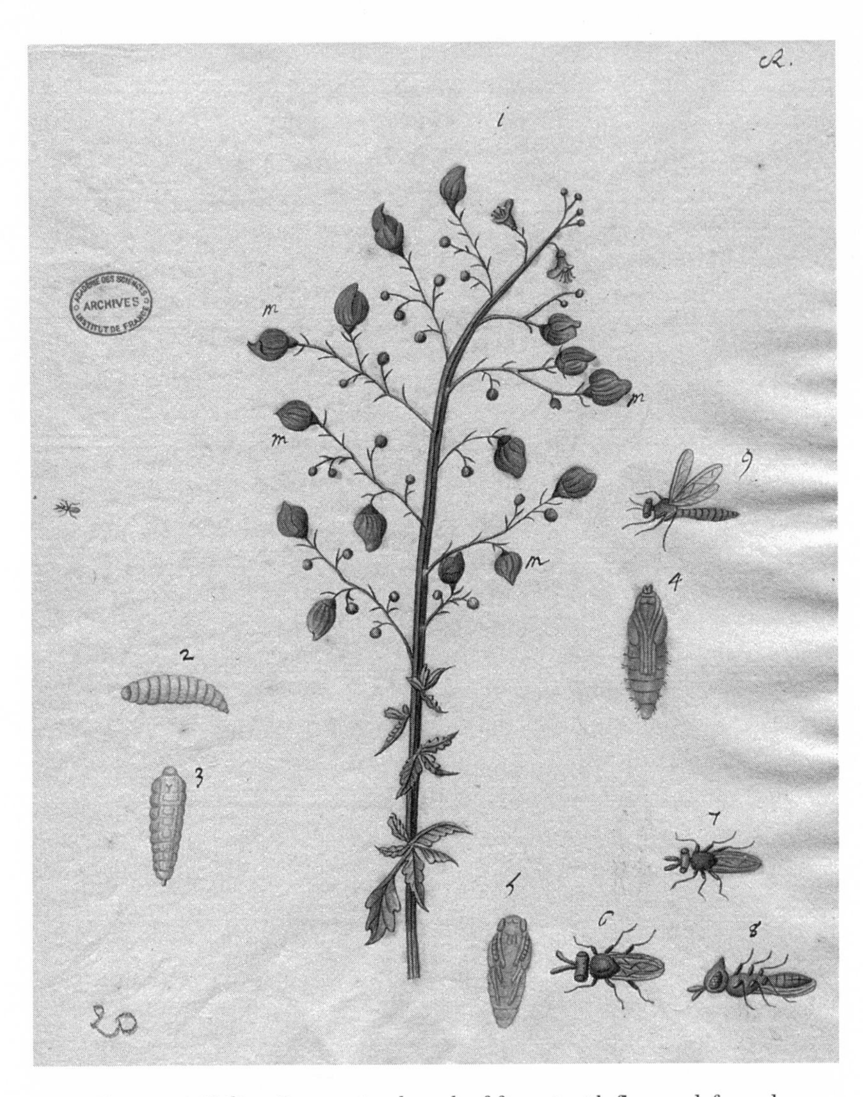

Figure 4.1. Hélène Dumoustier, branch of figwort with flowers deformed by flies laying their eggs, with various views of the insects found on the plant. "Found in 1742 at the beginning of September, on the levee of the Loire" (en route to vacation in Poitou). Archives, Académie des sciences, Fonds Réaumur, dossier 41. By permission of Archives de l'Académie des sciences, Institut de France.

✲

reporting, as Réaumur pointed out. The salient feature of the *mémoire* genre, for his purposes, was the room it afforded for first-person narratives of experiences, methods, and contingencies.

The author's literary choices pulled in readers at all levels of expertise; the books were "not only for ordinary readers, who seek only to amuse themselves with the surface of things, but equally for natural philosophers [*Physiciens*], who wish to go deeper."[9] Réaumur was asking his readers to look for themselves at the insects in their gardens and fields; he intimated that they would find that nature looked different when they trained their attention on previously unnoticed or even unappealing sights. Growths on tree trunks, which might have been seen as defects, become the homes of insects worthy of inspection. Even ugly objects become attractive: "Someone who would have looked with nothing but disgust at curled-up honeysuckle leaves, misshapen and dirty, in his garden, will perhaps see these same leaves with more pleasure than the cleanest and healthiest ones" after reading about the insects living inside the rolled-up leaves.[10] Instead of recoiling, the transformed viewer/reader will want to search out the insect that rolled up the edge of the leaf to form its cocoon, and even to follow it through its life cycle.

Readers did indeed see nature differently after reading these books. Pierre Billatte, a merchant in Bordeaux, borrowed the volumes from his friend Raoul (who collected the pine caterpillars discussed in chapter 2). Billatte wrote to say that he had been "enlightened by your researches"; his reading had prompted him to take his loupe and even his microscope outdoors to see what he could find: "The countryside, which previously offered only the pleasures common to everyone, on which we barely reflect . . . is now an inexhaustible treasure for me."[11] In the course of his walks, he happened on a pair of mating mosquitoes, and he hoped to be able to clarify some details Réaumur had not been able to see. Billatte had discovered the coupled insects on a chestnut tree, and he recounted his attempts to get close enough to see them with his loupe. In a second letter, he confirmed his identification of the insects according to the description he had read in Réaumur's book. "I willingly pardon their species for the stings that I have suffered from them, since they have procured for me the honor of writing to you, and of receiving such a gracious letter from you."[12]

As an academician, Réaumur used the royal printing house, the same presses that produced the academy's publications. He chose a format reminiscent of the *Histoire et Mémoires*, the annual record of the academy's work, with a slightly larger typeface and more extensive illustrations. Each volume opened with an essay more self-consciously literary than the narrative chapters. These essays gave the author the opportunity for general reflections on method, style,

Figure 4.2. J.-B. Haussard, bees observed in their hives (Charenton), with gardener moving a swarm. Réaumur, *Mémoires pour servir à l'histoire des insectes*, vol. 5, vignette. Courtesy of History and Special Collections Division, Louise M. Darling Biomedical Library, UCLA.

◆

and organization. He might also give advice about how to read and use the books, or he might use this space to respond to readers, update observations from previous volumes, or introduce the species to be discussed in detail in the following *mémoires*. The individual chapters, with their detailed histories of insect anatomy, physiology and behavior, often lapsed into narratives of observations and experiments, with dates and places and names of participants.

The first volume started with a long preface on "the natural history of insects in general" and continued for another 650 pages with chapters (*mémoires*) on caterpillars in general, their anatomical parts, the mechanics of their various maneuvers throughout their life cycle, chrysalises in general, and finally the emergence of the butterfly from the chrysalis. Even this was not enough space to cover all Réaumur wanted to say "in general"—though the supposedly general discussion was studded with detailed particular observations and suggestions about how to categorize caterpillars in easy and intuitive ways. He hoped to finish the overview of caterpillars and butterflies in the first volume, but it overflowed into the second—which didn't appear for another two years. Only part way through the second volume did the reader come to chapters devoted to individual species or classes of insects.

For all his attention to the behavior of caterpillars, chrysalises (pupae), and butterflies (including those now called moths), Réaumur did not pause for long over the problem of how to classify this profusion of species. Though readers might have expected "a detailed plan of the order in which we thought we ought to put our different classes of little animals," he admitted that he had given this up as too cumbersome.[13] So although he did name broad categories of insects and gave a rough classification according to easily observable characters, like number of legs, whether they live in groups or singly, or nocturnal versus diurnal habits, he announced at the outset his preference for full histories over the kind of structural or anatomical descriptions that would lend themselves to cut-and-dried taxonomic rules. Above all, he wanted to avoid describing each species simply for the purposes of classification; instead, descriptions would make up part of the full histories of different kinds of insects.

This choice to forgo a fully articulated classificatory scheme was not always to the taste of his readers, at least some of whom were hungry for an insect taxonomy. Even before the first volume appeared, he heard from a Montpellier botanist, Pierre-Augustin Boissier de Sauvages, about this. Sauvages indulged a lively interest in local insects and had recently presented to the Montpellier Academy a sketch of a new classification scheme for insects. When he heard about Réaumur's still-unpublished work, he sent his own "method" for classifying insects off to Paris, hoping for advice on improving it. "This is really the most curious and the most neglected part of the natural history of animals," he wrote. "I would be thrilled if you would make this subject as clear as Mr. de Tournefort has done for plants. I have no doubt that you will follow a similar or better order [than mine]."[14] When Réaumur confessed that he had given up on producing a comprehensive classification system, in part because of the vastness of the insect kingdom, Sauvages replied, "the reasons you give for not arranging your insects according to a studied and scientific method seem very thin, and who will be able to do it if you do not?" After all, Tournefort had proceeded with his flora without knowing all the plant species from other parts of the world that later would be successfully incorporated into his system:

> Without method, one cannot name the least insect properly, nor describe them except with confusion and at great length. . . . [I]f the insects pass in review before such a great inspector as yourself without being reformed, arranged, and put into order, they will not have the honor of being known for a long time to come, few people being able to produce large volumes of observations without a general and concise method to distinguish them.[15]

If Réaumur found it daunting to parse nature into categories, he found it almost as difficult to organize his own observations into an orderly sequence. Since caterpillars change into butterflies, the very objects of scrutiny do not remain stable; even worse, the most obvious traits for categorizing caterpillars do not allow them to be matched consistently with the corresponding butterflies. Nevertheless, the book needed some organizing principles. He chose certain easily recognizable traits and flexible categories such as "caterpillars that hang upside down vertically to metamorphose," "moths or nocturnal butterflies," "hairy caterpillars," "social insects," and "solitary insects." These pragmatic and vernacular categories make possible rough identifications even of species not included in the book. The reader naturally would like to know which caterpillar becomes which butterfly or moth. But grouping caterpillars and butterflies by the most striking characters, while apparently the most natural procedure, would only lead to confusion, as similar caterpillars do not all turn into similar butterflies. In short, if caterpillars were to appear together with their butterflies, "either the butterflies or the caterpillars would not be in proper order."[16] In the interests of connecting caterpillars to their butterflies, Réaumur left aside the abstract problem of order in favor of narratives that led readers first to the illustrative plates, and then out into nature to look for themselves. Réaumur never drew up a chart showing how classes related to each other, as was common in works about plants or shells, although he suggested that the reader might do this for himself:

> It is true that in this way the butterfly is often out of place, that is to say that it is not always with those that look similar. But if one wishes to correct this disarrangement, he has only to draw up a table of all the butterflies mentioned in the work, where they can be named in their true order, and where the plates with their figures can be cited. In this way he would see again all the butterflies appearing in the proper arrangement, he would see at the same time all the caterpillars from which they are born, and this would be a short and useful recapitulation of what he would have already read.[17]

Réaumur wanted his books to be useful to people like Billatte, who enjoyed identifying insects encountered in the gardens, fields, and forests. But he did not think this kind of quotidian activity warranted a full-blown classification table. "After all," he decided, "it is perhaps not as essential here to put the facts into a good order as to gather enough of those that merit attention. Perhaps any order is good, provided that we give detailed histories of each principal kind of insect."[18] Once good histories are written, they can be organized according to different schemes depending on the purpose: Valisnieri's scheme would work for finding insects in the countryside; Swammerdam could be used for understanding the transformations of different insects. To correlate these approaches, nothing more was required than to "draw up a table."

Leaving the construction of such tables as an exercise for the reader, Réaumur went on to what might be called the physics or the mechanics of insect life. At first glance, the table of contents of this initial volume gives the impression of a logical progression, from "insects in general" to "caterpillars in general" and on through chapters on the parts of the caterpillar, various kinds of butterfly, the chrysalis "in general," and the various steps of metamorphosis. In each chapter, the reader encounters rather an eclectic mix of specific observations, suggestions for further research, and topics that spin off from the core subject. Take, for example, the third chapter, "On the different parts of the caterpillar." It covers all of the anatomical structures in some detail, elaborating on new discoveries (e.g., light-sensitive "eyes" in some species), on various experiments, and especially on the silk-producing organs common to many species. The essay drew extensively on Malpighi's canonical work on silkworm anatomy—"a tissue of discoveries," Réaumur called it—and even reproduced the Italian anatomist's images of the dissected caterpillar.[19] The silkworm had been intensively studied by Malpighi and others because of its obvious economic interest, and it became here a prototype caterpillar. Moving from this example to caterpillars more generally, Réaumur explored the possible uses of the silk of other species. Examining cocoons of the "common" or "liveried" caterpillar, he noticed that the strength of the fiber depended on what the caterpillar was eating. "This difference in quality of the silks of caterpillars of the same species, which live on different kinds of leaves, undoubtedly comes from the different quality of the leaves they eat; this should push us to test whether we cannot put these caterpillars to work usefully for us, by feeding them certain leaves."[20] This suggestion hovers seductively as a promising avenue of investigation, to be followed up in the future, perhaps even by the reader.

The properties of materials were as much a part of the natural history of insects as their feeding habits or the changing of their skin. Silk as potentially usable thread was one thing, but what about the properties of the sticky liquid from which it was made? This had qualities "equally curious and useful" but had never been adequately investigated, either by "lovers of physics [or] those who love the arts." Réaumur himself had been interested in the stuff for more than twenty years, at least since his work on spider silk, but he had never pursued a serious investigation of silk from other insects. "Perhaps I will find myself [one day] in circumstances that allow me to explore this; but I exhort those who are masters of their own time and willingly employ it for experiments, to experiment on any ideas that this liquid might inspire."[21] Even without systematic trials along these lines, Réaumur noticed that the substance of the caterpillar's threads could function as a kind of varnish—spread over a surface, it would form a sheen much like a lacquer. He had seen this in the jars where he raised caterpillars; when they spun their cocoons right up against the glass, the shells left a glossy residue when they were detached, a layer "as uniform and as shiny as the glass itself." With no vestige of the threads visible, the material had formed into a skin, about the thickness of a sheet of paper.[22] Could this substance be put to practical use as a varnish? The question was, he suggested, at least worthy of investigation, especially since these caterpillars could sometimes be collected in huge numbers. "Who knows if something that is useless to us today will not become useful some day?"[23] This was not meant idly; he was asking his readers to take such possibilities seriously and to apply his methods to curious phenomena, however small and apparently insignificant.

If the silk of caterpillars seemed potentially useful, their transformations were surely among the most curious events in the insect world. Turning to the chapters on chrysalises and the complex maneuvers employed in metamorphosis, the reader came upon not only tips for how to observe these ephemeral events but also an interpretation of what was going on when the insect changed form. The sequence of narrative and image supported a theory of preexisting germs and of species stability. Malpighi had found the butterfly's eggs in the chrysalis of a silkworm; Réaumur worked with a larger species (*chenille aux oreilles*, or gypsy moth) and found recognizable eggs in the caterpillar itself. Although the eggs were too small to reveal future caterpillars inside them, the implication was clear: all organized beings must have existed since the time when all creatures were formed; they become visible when favorable circumstances allow them to unfold and develop. This conviction is rooted in observation but also on a commitment to the possibility of reasonable explanation, or what we might call the possibility of scientific knowledge. If each insect were really a new formation or creation, "we would have to renounce explanations of how they form." The notion of simple development of preexisting structures "allows us to make observations that can at least give us an understanding of the order in which they happen."[24]

Metamorphosis gave Réaumur the opportunity to articulate a rational approach to the study of apparent marvels. "What reasonable idea could one have" about the sudden change from caterpillar to chrysalis and chrysalis to butterfly? Before the relatively recent progress made by what he called the "new philosophy," these changes had been seen as spontaneous transformations, and inherently inexplicable. Anyone foolish enough to suppose that "a little rotten flesh, or rotten wood, could become the legs, wings, proboscis, eyes, in a word, the whole body of an insect, made up of such admirable organs, of so many muscles, nerves, veins, arteries," would have no trouble believing that some bits of the chrysalis could spontaneously turn into the butterfly's wings, or that the twelve legs of the caterpillar could somehow provide material for the six legs of the butterfly. In reality, Réaumur explained, such abrupt and complete transformations are "as chimerical as those of the fables."[25] The anatomists of the previous century had discovered this when they dissected insects and saw the distinctly recognizable parts of each form—caterpillar, chrysalis, butterfly, eggs—encased in layered sheaths. These careful anatomists, with their microscopes and their scalpels, had destroyed the "false marvel" of metamorphosis, which could never be more than a confused idea, "but at the same time they left us plenty of real marvels to observe."[26] Following Swammerdam and Malpighi, Réaumur argued that the chrysalis was nothing other than a butterfly with its parts hidden beneath an outer shell; the caterpillar literally contained the butterfly, including the eggs, and thus the future generations of caterpillars as well. On this view, the parts of the butterfly grow and develop within the caterpillar, just as all other organized bodies develop, through the expansion and solidification of the parts. The caterpillar spends its life eating and digesting leaves to feed the butterfly, "just as mothers prepare the food that is transported to the fetus [in utero]."[27]

The precise and "philosophical" meaning of metamorphosis took the phenomenon out of the realm of the fabulous and into the mechanical world of physical science. And indeed Réaumur and his assistants approached their insect subjects with the tools of experimental physics—air pumps, thermometers, scales for measuring changes in weight, and distilling apparatus. Numerous experiments with immersing caterpillars and chrysalises in oil and water for varying lengths of time led to the conclusion that air passes through the stigmata of the chrysalis but not of the caterpillar. Respiration thus works differently in the different life stages, a result confirmed by the air pump. "I put chrysalises of different species in a small receiver [attached to the air pump]. I was curious to see what would happen to the volume of their bodies." The outer skin of the chrysalises seemed ill suited either to let air out or to stretch. As the air was pumped out of the receiver, the volume of each little body increased, but instead of inflating evenly like a balloon the compacted rings unfolded along the length of the body, "forc[ing] the body to extend in the direction where there was the least impediment." Caterpillars, on the other hand, did not expand under similar circumstances, because the expanding air could escape through their skin. "Finally I put under the receiver [of the air pump] a glass vase containing water that had been purged of air, and in this vase I kept a chrysalis submerged in the water. I had encircled its body with a thread, and attached a weight to the ends of this thread. After one or two strokes of the piston, large bubbles of air appeared on each stigma; they came out in jets, and there were only a few small bubbles from other parts of the body of the chrysalis."[28] Here the initial observations of anatomical differences between chrysalis and caterpillar led to a battery of experiments to clarify the physiological process of respiration.

The minutiae of the living insects' maneuvers and mechanisms fascinated Réaumur endlessly, and he counted on eliciting a similar response from his readers. He filled four chapters with observations of caterpillars' various means of positioning, hanging, securing, and protecting themselves. "All caterpillars act as if they knew what was coming next; but different species use different means to prepare themselves for this metamorphosis."[29] These various means, the products of the caterpillar's skill or industry, were among nature's "real marvels." Even with the end result plainly visible—a shell made of silk and hairs, or a suspension system used first by the caterpillar and then by the chrysalis—it was never obvious just how the insect had contrived to make it. The message for the reader was clear: renounce fruitless guesses and get on with the task of contriving to see the actions of feet and jaws and silk glands. "It is only by seeing these insects operate that one can discover their mysteries, but it is difficult to seize the moments when they can be seen in operation."[30] The transformation of a caterpillar was at once utterly mundane and frustratingly elusive. And it had never been fully described, even by Malpighi and Redi, who had explicitly mentioned seeing it.

However laden with detail, Réaumur's accounts left open many possible avenues for corrections, additions, and novel ways of seeing into the lives of insects. Making a virtue of necessity, he admitted to leaving many things undone and unseen, but he challenged his readers "to dig even further into one of the most curious subjects in physical science."[31] In the preface to his second volume he continued to encourage readers to see for themselves:

> As more attention is devoted to insects, people will make observations that have escaped me. Even those I report are sometimes imperfect; sometimes it happens that I discuss a caterpillar for which I have not yet obtained the butterfly, and discuss a butterfly whose caterpillar is not yet known to me. I say this to alert observers to all that remains to be done, and to invite them to profit from any occasions that will show them the whole of what I have only seen halfway.[32]

SEEING THE EPHEMERAL

Réaumur contrived to see the emergence of the chrysalis from the caterpillar many times, and to describe it in detail, by bringing as many of the insects as possible into his workroom, where he could keep them under surveillance (see fig. 3.1). Rather than simply abstracting his description from the notes of his observations, he included narratives of circumstances and procedures to give his reader a taste of the strategies and contingencies that went into making natural historical knowledge. These stories were meant to inspire admiration for his ingenuity and patience, no doubt, but also to provoke similar observational adventures that might well result in further corroborative detail. To frame his report on the transformation from caterpillar to chrysalis, for example, Réaumur recounted the circumstances that made possible a large-scale investigation of a problem that had so often eluded the gaze of naturalists. In July 1731, an extraordinarily heavy infestation of gypsy-moth caterpillars in the oak trees of the Bois de Boulogne stripped the trees of their dense canopy of leaves. When the oak leaves had been eaten, the insects were ready to metamorphose, and Réaumur had hundreds of them gathered and brought into his home. There were so many of them spread out on a large worktable in the Hôtel d'Uzès that hardly fifteen minutes would pass without one of them breaking out of its skin.[33] The event took only a minute or so; the large numbers meant that a normally rare occurrence happened over and over again under conditions of excellent visibility. He found he could interrupt the process by dropping an insect into spirit of wine while it was shedding its skin; there it would continue its motions, briefly, before dying part way through the transformation. These insects, immobilized at various stages, contributed to completing the picture as the fleeting movements were frozen and preserved. The artist could draw the transitional stage, and the outer layers could be separated with dissecting tools to reveal the parts within. The final text narrates the sequence of motions and actions, portrayed statically in the illustrations. Once you have seen it, he claimed, pupation was "easy" to observe again.[34] On the one hand, Réaumur presented his detailed descriptions as novel discoveries; on the other, he explained how anyone could do the same, by gathering enough caterpillars "to multiply at will events which otherwise would be very rare, to a point where it is almost impossible to miss seeing the event."[35]

Each of Réaumur's chapters incorporates little scenes of the observing life, framed with details that put these scenes into particular places, either in the home, outside in the garden or the woods, or even at the academy. For

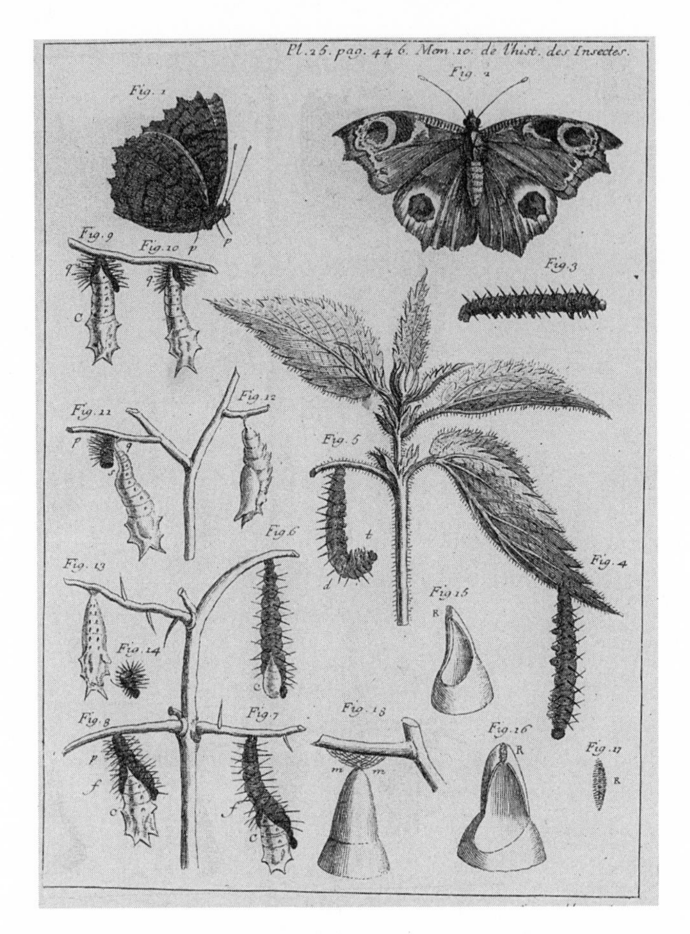

Figure 4.3. Philippe Simonneau, stages of transformation from caterpillar to chrysalis, with butterfly. Réaumur, *Mémoires pour servir à l'histoire des insectes*, vol. 1, plate 25. Wellcome Library, London.

✻

example, in a chapter on the construction of the rounded cocoons spun by some species to protect the chrysalis, we learn about a flourishing colony of tiny caterpillars living on an east-facing wall outside the room in the Louvre where the Academy of Sciences held its meetings.[36] They were first noticed in the month of May by P.-L. M. de Maupertuis, a fellow academician, who brought Réaumur out to the terrace to see them; subsequently the two men made it a point to visit the caterpillars every Wednesday and Saturday, when the academy met. No doubt they welcomed the chance to stretch their legs and encounter a bit of nature after the stilted performance of scholarly duty—reading papers aloud to each other—inside the room. The insects belonged to an unknown species, notable for barely moving and for subsisting on the nearly invisible plants growing in the cracks of the stone; their numbers fluctuated from week to week. "You can well imagine that I did not content myself with observing them on the wall where they were thriving. I took some of them home and enclosed them in jars."[37] He put pieces of lichen-covered stone in the jars with them and noticed that they were soon stripped of the plant matter (though he never actually saw the caterpillars eating it). After about six weeks in their jars they started to construct cocoons, anchoring them to little fragments of stone. They pulled out their own bristles and "planted them like palisades" as protection for the chrysalis, covering this structure with a very thin layer of white silk. The butterfly, when it emerged a few weeks later, was small and nondescript. Réaumur could have simply described the few distinctive features of this insect—its appetite for lichens and its peculiar habit of incorporating its own bristles into its cocoon. Instead, he embedded these details in the tale of how the caterpillars, living among the lichens clinging to the cracks and crevasses in the walls of the best address in Paris, had attracted the attention of an observant academician, before being taken into captivity for closer observation. Following this little story, readers could see that no insect was too trivial to be worthy of the naturalist's gaze, however grand the location of its habitat or the credentials of its observers.

Although Réaumur could be excruciatingly longwinded, the nonlinear trajectory of his *mémoires* captured the ins and outs of his reasoning process as well as his observational tactics; he wove into the story of his successes and failures additional details noticed by other people and the stories of odd behaviors of the insects themselves. Readers could see how these disparate elements all came into play, as they relived the circumstances and strategies of specific experiments and particular insects on the way to considering "general" questions, such as what happens in the chrysalis before the butterfly emerges, or the length of each stage of the metamorphosis. So, for example, having established through a series of experiments the "simple principle" that the chrysalis gives off moisture as its structures solidify into the butterfly's parts, and that the rate of transpiration depends on the temperature, he decided to test whether he could speed up the process by the application of heat. Or, noticing a wasp chewing on his windowsill in Charenton, he resolved to investigate each step in its process of transforming the wood fiber into a paper nest.[38] Such examples could be multiplied almost indefinitely. Without reporting every experiment in full detail, Réaumur generally gave enough information about circumstances, techniques, and tools that the reader could adapt them to other locations.

READERS RESPOND

Readers in different places, with different resources, might develop their own versions of such investigations, as we have seen in the brief example of Billatte's observations of mating mosquitoes. Such readers, like Billatte, often engaged with the text directly, in the course of their own sustained researches; that is to say, reading and observing went hand in hand. In some cases, these fellow observers were known to Réaumur, and their responses simply extended ongoing conversations and correspondence. Others approached him after reading his books and following up on suggestions they found there. His old friend Bazin was in some ways the ideal reader. He responded enthusiastically as soon as he had finished the first volume: "I do not know which to admire the most," he wrote, "the industry of the insects or your own industry in revealing to us the imperceptible mysteries where nature seemed to be hidden beneath veils so much more difficult to pull away because they are almost impalpable." The ingenuity and skill of the insects is matched by those of their chronicler, as he works first to reveal their secrets, and then to recount his efforts for his public. Bazin captures nicely the double register I have been examining, the maneuvers of the insects paralleling the maneuvers of the naturalist. He joked that readers unfamiliar with the peculiarities of insect behavior might have trouble suspending their disbelief when faced with the sometimes startling phenomena in the book's pages. They might not believe that "you have written not a fictional romance about nature but the true facts and movements of insects."[39]

Bazin applauded not only the wealth of details, but also the sustained effort and ingenuity they represented. His intimate familiarity with this kind of work, and his many contributions to Réaumur's enterprise, meant that he could express his reservations and criticisms as well. On the matter of the relation of caterpillar to butterfly, for example, Bazin objected to Réaumur's analogy:

> You say that the caterpillar is a mother who feeds the butterfly as her fetus. This proposition seemed to me hard to digest. It supposes that the egg of the butterfly would contain two animals encased one inside the other, of which the progenitor would turn in twenty-four hours into the substance of the generated, at the time of its metamorphosis. Besides, in [your] final lines . . . it seems that you are making the caterpillar the fetus of the butterfly [rather than the other way round].[40]

Most of Bazin's comments, though, address matters of style or small inaccuracies in the descriptions and illustrations, betraying the intimacy of his engagement with text and images. Since he was perfecting his dissection technique on the heads and tongues of caterpillars just at the time he received the book, Bazin noticed that Réaumur's descriptions, and especially Simonneau's illustrations, did not always match exactly what he was seeing through his own magnifying lens.[41] Bazin was evidently dissecting and reading in parallel: "Seeing that there was no way to get help from people around here on anatomical matters, I got it into my head to do it myself with the help of your book." Shortly thereafter, he sent his own drawings of dissected heads to correct the printed illustrations (see fig. 2.2).[42] In another instance, he corrected Réaumur's unequivocal statement that butterflies do not retain visible stigmata: "In handling a female silkworm moth, I was surprised to see, in spite of the fuzz covering it, perfectly visible stigmata on the rings [of the thorax], just as they were on the chrysalis."[43] And he went on to search for the analogous structures in other kinds of butterfly.

Bazin used Réaumur's books as references and guides; he also worked through them methodically and made an extract of each chapter, drawing up an "analytic table" or index for each volume. His comments about these tables give us some insight into his reading practices, how he used the books and how he thought others might use them. He conceived the abstracts initially for his own use but then offered them to the author for possible inclusion in a second edition.[44] When he received the second volume in June 1736, he read through it once "with as much pleasure and contentment as the first," and then read it a second time "with more attention" in order to make the abstract. The second, critical reading also produced "remarks and objections," usually corrections to fine points of anatomy.[45] Five months after receiving the second volume, he reported, "Now we are in the dead season, and I will take advantage of it to devote myself entirely to the table for the second volume. It seemed to me that I would do better to make a single table for the two first volumes. When the second is finished, I will reduce them into one." Eventually, Bazin sent a copy of the two analytic tables to Paris, and he repeated the task for subsequent volumes as they came out.[46]

However engrossing the exploits of caterpillars might be for those who watch them habitually, Bazin was afraid that most readers would find the large books cumbersome, and made specific suggestions for an abridged second edition that would appeal to a more casual readership and make the plates and descriptions easier to use.[47] He reported that some readers in Strasbourg had difficulty sticking with the detailed accounts: "Many people have borrowed your book from me—Jesuits, men of letters, and others—and none of them got past the third chapter. The most stubborn skimmed the rest by jumping around, and that is because of the organization. I tell you this with regard to the next volume."[48] Réaumur took some of this to heart and kept the next volume shorter. He acknowledged in print that some readers might expect the chapters to be more amusing than they would turn out to be: "It is certain at least that somewhat extended extracts of these papers would please many readers more than the papers themselves."[49] With a denser concentration of singular and noteworthy facts and fewer of the "dry" details that filled the descriptions, such extracts might be more widely appreciated, but he did not propose to edit out the details. In any case, he had already advised those reading only for amusement to skip the long sections in the first volume where he defined the characters of different categories of caterpillars and butterflies. On the other hand, anyone who wanted more than superficial knowledge would be pleased with "order, method, and exact details." Such serious readers could be found in unexpected places, as is evident in the story of the Duchess of Maine's visit to his insect collection—"those I keep alive, as well as those I preserve dead." The duchess demonstrated by her pointed questions about the collection's organization that "not only had she read these unprepossessing *mémoires*, but that all the characters of the classes and genres set out in them were even more present to her mind than to my own."[50]

The hefty first volume may have been daunting to the casual reader, as Bazin suggested, but plenty of people nevertheless responded enthusiastically as the subsequent volumes appeared. Their letters often arrived with specimens or queries, as they answered the author's call to "pay more attention to insects" and to accumulate new observations that had escaped him. Bazin was, as we might expect, one of the most articulate about his reading experience:

> I devoured all in one breath the fourth volume of your *Mémoires*: I will only say that after my reading was finished, and having rested my mind a bit, I found myself seized with a ravenous hunger, and an extreme need to sleep. However, I had had along the way quite a tasty meal, I had savored at leisure the praises with which you were so kind as to honor me. . . . By the most singular chance in the world, I was just working on those two stigmata, and your prediction about them, when someone brought me a butterfly. I thought that fate, favoring me in such a fashion, wished to reply to your promise. I started therefore at once to search for these two stigmata.[51]

The physical book—text and images—connected Bazin in his study in Strasbourg to Réaumur and Dumoustier in Paris.

Another reader who found the books engaging was Pierre Baux, a provincial physician in Nîmes who spent his leisure time in a variety of natural history pursuits, from keeping temperature records to collecting birds and insects. Lacking, as he complained, sufficient time to spend observing nature for himself, Baux eagerly anticipated each of Réaumur's books:

> I always read the previous volume four or five times, before the new one appears. It is such a lovely story that one devours the book when one reads it for the first time. I do not tire of reading it, but it annoys me that the printers make us wait such a long time. If I wanted to say everything I think about this excellent book, I would run out of paper.[52]

Even discounting the obvious flattery, Baux's testimonial shows that these books were read and reread.

Many readers used the books intensively, as we have seen in the case of Réaumur's friend and contemporary Bazin. The next generation of naturalists also drew inspiration from the histories of insects. In his autobiography, Charles Bonnet recalled his first encounter, as a precocious sixteen-year-old in Geneva, with one of Réaumur's books. Bonnet had read Pluche, and he had started to notice the insects around him. But it was not until he read Réaumur that he devoted himself seriously to natural history. Bonnet described coming across the first volume on insects by chance on his philosophy professor's table. He only had time to "devour" a few pages before being called back to his lesson, but the effect was powerful nevertheless. "I was seized with astonishment and joy; . . . I was enchanted to see these insects, so despised by the vulgar . . . in some way ennobled by the pen of the academician." When he asked to borrow the book, the professor told him to read Pluche instead. "I have read and reread it," Bonnet replied. "I know it almost by heart." No matter; Réaumur was too learned for the boy, his teacher insisted, and besides, it was a library book. The young man let a few weeks go by before going to the city library himself to find the book, seeking out the librarian he thought would be most likely to grant his request. "Go away," the librarian told him. "We do not loan such books to young people." Only after returning again to beg for the librarian's indulgence did Bonnet acquire the coveted volume, "the book . . . that would make a naturalist of me." He read "day and night," making long extracts—never imagining at the time "that I would one day be the owner of this great work and that I would receive it from the very hand of the illustrious author."[53] The anecdote, though Bonnet told it as an almost mythical quest leading to a conversion experience, captures the visceral thrill experienced by the young reader. In fact, it was not long after this initial reading of Réaumur that Bonnet worked up the courage to write to Paris, and soon enough he was receiving leatherbound volumes as gifts. Réaumur used the books to reward the efforts of his young admirer, who provided him with many sustained and novel observations, as we shall see in the following chapter. Bonnet's experience, still vivid to his mind's eye years later, is a telling example of the connection between reading and the inculcation of habits of observation.

JESUITS AND GENERATION

Réaumur, like most of the observers and collectors he inspired, spent his time gathering and managing particulars rather than building theories or speculating about metaphysics. Nevertheless, the world of insects held lessons about questions like generation and the origin of organization, even if indirectly. He made clear at the outset that he was following in the footsteps of the experimentalists and microscopists of the previous century—Francesco Redi, Marcello Malpighi, Jan Swammerdam—some of whom had designed experiments to debunk the ancient notion of spontaneous or equivocal generation. The Aristotelian cycle of generation and corruption, with the lowliest creatures (insects, worms) emerging spontaneously from rotting matter or mud, had lost credibility in the face of mechanical philosophy and experiments like those of Redi, who showed that maggots emerged from eggs laid by flies, rather than from the rotting meat where they were found. By the eighteenth century, the insect dissections of Malpighi and Swammerdam and Redi's experiments on the generation of maggots were widely hailed as emblems of the empirical attitude at the heart of the "new science," with its repugnance for speculation and rejection of the absolute authority of ancient texts.[54]

But some ancient beliefs died hard. Reflecting on the tenacity of assumptions about the connection between insects and decaying matter, Réaumur pointed to the learned Jesuits Athanasius Kircher and Filippo Bonanni, who had held onto the "absurd sentiment" of spontaneous generation in spite of experimental evidence to the contrary.[55] "It is truly surprising," Réaumur remarked, "that such ideas could survive even after people had started to look at the smallest insects with philosophical eyes."[56] The brief for unequivocal generation—animals universally generate offspring of their own species—was a proclamation of the stability, consistency, and intelligibility of nature, and Réaumur listed this as one of the fundamental "principles of the natural history of insects."[57] "Philosophical eyes," which were not the eyes of stubborn Jesuits, would not see miracles, fables, or randomness in any part of nature's three kingdoms.

While Réaumur's first volume was generally well received in the press, the Jesuit *Journal de Trevoux* took umbrage at the condemnation of Kircher and Bonanni for their commitment to equivocal generation. After praising Réaumur's observations as "very detailed, very ingeniously handled, and very curious for anatomists and for amateurs of anatomy and of good physics," the reviewer came to the defense of his brothers on the matter of equivocal generation.[58] Kircher's idiosyncratic system based on a plastic seminal virtue spread throughout living things "might not be true," the reviewer admitted, "but it is very ingenious, and besides, completely consistent with the kind of chance that gives birth to so many insects, and to the quite marked association of certain insects to certain bodies."[59] As Réaumur reported to a correspondent in Switzerland:

> These fathers [the Trevoux journalists] do not like the academy or anyone who has anything to do with it . . . . They have put forward several propositions that I will address in [my] second volume. For example, they do not think it has yet been sufficiently proven that no insect is ever born from rotting matter. . . . And plenty of other propositions just as ridiculous, from which one can draw strange consequences.[60]

Indeed, the way his Jesuit critic took for granted the random origin of insects could hardly pass without comment, and Réaumur responded pointedly in his next volume, published in 1736. How, he asked, could the learned journalists of the *Journal de Trevoux* hold such benighted views after the observations of Swammerdam, Malpighi, and others "ought to have opened the eyes" of anyone who still believed in spontaneous generation?[61] If insects appeared spontaneously, with no rhyme or reason, systematic study of insect life cycles would be a pointless exercise. "As long as one believed that [insects] came from rotting matter, the most curious part of their history, everything that touches on the way they perpetuate themselves, did not seem to be worth studying."[62] To get beyond the superficial, and to avoid jumping to unwarranted conclusions, observers had to notice all relevant circumstances—especially the presence of egg-laying insects near rotting matter. If we pay attention, Réaumur argued, "we will be convinced that chance has no more part in it than it has for large animals, or for man himself."[63]

In rejecting the association of insects with chance, Réaumur articulated a fundamental principle that he had earlier taken almost for granted, "that insects of all species produce individuals of their own kind."[64] Insects, after all, were just another kind of animal—this was a guiding premise of the natural history of insects. And why would God have given insects organs of generation if they were not to be used?[65]

> If all that was required to give birth to ordinary flies, to bees, to wasps, to beetles, to frogs, etc., were a little rotten meat or some plants reduced to manure, there would be no need for the two sexes found in all these different species of little animals. The parts essential to the growth and fertilization of eggs would be superfluous.[66]

Réaumur's attack on Kircher opened out into a defense of his own method, where observation accompanied by critical reasoning would distinguish credible from equivocal facts. There was no lack of experimental evidence, going back to Redi, who had seen the eggs laid by flies and mites hatch into the insects supposed to arise from rotting matter. Events as elusive as the mating of fleas, and anatomical structures as tiny as the sex organs of cheese mites, had been witnessed and documented by "attentive and enlightened observers." Such observations (some of which he had confirmed himself, and which he invited his readers to see in their turn) Réaumur regarded as unequivocal, verging on a "physical demonstration."[67] Kircher's "excessive love of the marvelous" and his strong predilection for his own system made him an observer not worthy of credit. Réaumur's theory of species fixity, tied firmly to the preexistence of germs, was rooted in both empirical observation and admiration for divine providence.[68] Or perhaps it would be more accurate to say that a long and patient practice of observation "with enlightened and philosophical eyes" did nothing to diminish his convictions about the wisdom of the divine designer. Réaumur, like most of his readers, saw no contradiction between providentialism and mechanical explanations, and he pushed hard to reveal the mechanisms behind the operations of insects.

Some of the differences separating Réaumur from his Jesuit critics simply reflected their different priorities: the Jesuits were concerned with piety and Christian dogma (especially miracles), whereas the naturalist was concerned to inculcate critical habits of mind and attentive use of the senses in his readers. As each new volume appeared, the Jesuit reviewers pounced on any reference to a person or notion that might be construed as infringing on what they construed as theological territory. A critical reference by Réaumur to Malebranche's use of the emerging butterfly as an image of resurrection was one such red flag.[69] The reviewer noted that long before Malebranche the church fathers had used the language of metamorphosis in connection with resurrection, and that the chrysalis could legitimately be taken as a figure or image "appropriate to give us an idea of a power capable of performing the greatest miracles, and in particular the miracle of our resurrection."[70] Réaumur responded that Malebranche should have known better, since he had kept and observed insects (specifically, ant lions) himself. Moderns, he implied, had no excuse for arguing like the church fathers, who had not had the benefit of the new science.

The sniping continued between the academician and the Jesuits with successive volumes. The *Journal de Trevoux* would give generally accurate extracts, interlarded with pointed comments that then provoked a response in the preface of Réaumur's next volume. When the "learned journalists" described the natural history of insects as "an object so vast that the life of the most laborious man hardly suffices to skim the surface," Réaumur replied sharply, "Because we cannot know everything . . . are we to condemn ourselves to complete ignorance?" Réaumur traced the hostility of the Trevoux reviewers to his exposure of the "puerility and absurdity" of Kircher's system, but then pointed to a longstanding and more general animus of Jesuits toward the Academy of Sciences:

> Everyone who reads the *Journal de Trevoux* knows that for many years the works of that academy, so respected by all the savants of Europe, and works bearing the name of some of its members, are treated in the *Journal* as they would be if the academy as a body had undertaken the critique of Father Kircher.[71]

Readers noticed, and even relished, the exchange of barbed remarks with the Jesuits. Bazin remarked that "the good fathers do not understand banter at all, and wish us to respect even their idiocies. I doubt that they will leave you without a response and I find you quite courageous to enter into combat against a society with thirty or forty thousand men to oppose you."[72] He looked forward to seeing the further defense of Kircher: "this little literary dispute will certainly give pleasure to spectators; luckily you have reason and laughter on your side."[73] Another correspondent declared himself "charmed" by the preface to the second volume, especially the refutation of spontaneous generation. "I do not know why the Jesuits are still infatuated by the false opinion of the origin of insects from rotting matter. . . . Do these gentlemen love the truth so little that they could voluntarily close their eyes to the evidence?"[74] Ironically, the Jesuit defense of spontaneous generation, however wrongheaded it looked to this correspondent, stimulated Réaumur to articulate his principle of the invariance of species, and the origin of all individuals from others of their own kind, much more explicitly than he would have done otherwise.

5

Natural Prodigies

ASEXUAL REPRODUCTION AND REGENERATION

Opposition to spontaneous or equivocal generation was a foundation stone of Réaumur's natural history, a position shared widely by his contemporaries but still somewhat controversial in the 1730s—as the exchange with the Jesuits attests. Insect observations spoke directly to the question by clarifying the chain of development from egg to intermediate forms (worms, nymphs, caterpillars, chrysalises) to adult and thence back to egg. The mechanics of fertilization—how it was accomplished, including the anatomical parts and the behavior associated with copulation—became an essential part of the "complete history" of any species. Contriving to see all the maneuvers associated with mating might well be as close as the observer could come to understanding generation.

Conception itself remained beyond the limits of sight, and hence beyond the reach of observation and experiment. "We must recognize," Réaumur admitted,

> that conception, the instant when generation begins, is the moment when an animal, an embryo of indefinitely small size, begins to grow; because if we want to see how germs are formed, if we want to imagine the means of producing the organization of the animal body either from liquids or from more solid matter, we will soon realize the impossibility of producing such a work [*ouvrage*] from the parts contained in the embryo when they are first evident to the senses; we could not find the necessary apparatus there to produce such astonishing machines.[1]

We humans cannot see how living beings—"such astonishing machines"—might be made from the available materials, but even God does not form animals at each act of generation. The origin of those "astonishing machines," lying beyond the reach of human imagination, can only be referred back to

God. The creation happened outside of natural history and does not, of course, need to be explained or even imagined. Bracketing off the creation of new life from the purview of the naturalist, or of natural science, left plenty of territory for exploration, filled with unsolved problems and a dizzying variety of modes of reproduction in the insect world.

Spying on the sex lives of insects led naturalists to consider how eggs were fertilized, laid, protected, and hatched, leaving aside unanswerable questions like the location of the preexisting germ or the trigger for expansion and development. At times, they went so far as to manipulate the mating process in order to see it better, or simply to see what would happen in contrived circumstances. Take Bazin's rather playful attempts to mate butterflies of different species with each other. He wanted to see whether insects, like larger animals, might be capable of producing monstrous, or hybrid, offspring. For these crosses he chose female gypsy-moth caterpillars, for their placidity, and male silkworms, whose behavior tended to be "the most lively in love." He raised both species in large numbers but had to manipulate the timing of their cycles artificially—the gypsy-moth caterpillars were already turning into chrysalises when the tiny silkworms first hatched from their eggs. He put the chrysalises on ice to slow their development, and when the silkworm cocoons were ready, he brought both kinds into the sun, hoping to get them to emerge simultaneously and then to mate. Unfortunately, the gypsy-moth caterpillars had suffered from the chill; the few moths that hatched were barely able to emerge from their shells, and "those that I helped [to hatch] were never able to recover enough heat to become amorous." Ever the optimist, Bazin found some replacement gypsy moths in a nearby field, though it was late in the season. He placed these near the male silk moths, which proceeded to act "as you know they are accustomed to do," but the females "stubbornly refused their advances." Bazin was disappointed, but decided that the results were inconclusive, since the wild females had probably already mated with their own kind.[2] Two summers later, he tried again, switching the sexes: "I mated male gypsy moths with female silk moths. These males spared no efforts to satisfy their urgency, [and] the females lent themselves to it with a will; but nothing happened from all that, no more than from the male silk moths and the female gypsy moths. Whatever accommodation they make, they were not at all made for each other." In a postscript he remarked that he had just noticed some eggs laid by a female silk moth enclosed in a jar with a male gypsy moth. The eggs turned purple and "seemed fertile." Hoping that he was not mistaken, he determined to guard the eggs carefully until they hatched the next year "to clear up this little mystery."[3] Note that Bazin, though he was not overly invested in the results of his crosses, did not perform them casually—they were part of his much larger program of cultivating and experimenting on local caterpillars and butterflies.

Bazin took the inspiration for his hybridization efforts from Réaumur's description of his butterfly observatory. To capture the elusive moment of hatching, which takes only a few minutes for each insect, Réaumur pinned dozens of gypsy-moth chrysalises to the tapestry on the walls, within view of his writing table. Whenever one of them showed signs of activity, he could put aside his pen and paper to watch. Upon emerging from their shells, the females moved sluggishly, but the males immediately and energetically sought out a potential mate, copulating as soon as they found one. Not content with watching, the naturalist got into the act:

> The ease with which I could pair off [literally, "marry"] these butterflies inspired me to do so several times. . . . I got a male into a glass jar; he kept on trying to escape, making all the motions of a lively and wild insect in such circumstances. While he was thus agitated, while he was flying everywhere in the jar, I took away my hand and quickly put the opening of the jar against my tapestry at a spot where there was a female moth. The male continued to flutter around and as soon as one of his haphazard flights brought him in contact with the female, immediately all movement in his wings stopped, and soon the mating was accomplished.[4]

The observer could see how the male lined himself up with female's body, but the wings hid the actual copulation. "The little mystery that the wings cover can be made accessible to the eyes if one causes these males to mate with females positioned against the sides of a glass jar, as I have done several times."[5]

Watching and manipulating the mating and reproduction of insects became a staple genre of observation. In the 1740s, several different types of insect became the focus of attention for their striking, even alarming, modes of reproduction. Hermaphroditic snails and worms, viviparous aphids, and even the regenerating freshwater hydra were not entirely unknown before 1740, but their most striking features had either not been noticed or had not been systematically investigated by earlier observers. Novel observations, many of them directly inspired by Réaumur's accounts of what he had seen, proliferated in this period and set off a kind of ripple effect throughout the natural history community. Some of these phenomena struck a chord as well with a wider public, beyond the core group of accomplished naturalists. Réaumur had spent years uncovering strange and sometimes surprising phenomena in gardens and forests, rendering the familiar exotic and exciting to his readers.[6]

When faced with unexpected modes of reproduction, like parthenogenesis and regeneration, naturalists responded by trying to make them intelligible and familiar, by repeating startling observations many times over, and by searching out additional species with the same peculiarities. Some of these things were initially regarded as prodigies; but they became the most mundane and replicable of prodigies through the hard work of their observers. Here I look at how some of these observations were managed, communicated, and ramified through personal connections and publications.

PARTHENOGENESIS

In 1724, Claude-Joseph Geoffroy took a plant louse from an elm gall, kept it in isolation in a jar, and in due course watched it give birth to live insects. He saw no fertilization take place and raised the possibility of parthenogenesis, or reproduction without mating.[7] There the matter rested, until Réaumur explored the behavior of many species of viviparous aphids in the third volume of his *Mémoires* (1737). As gardeners know all too well, aphids cluster in large numbers on the stems and buds of many kinds of plants, in full view of an attentive observer. Réaumur found he could watch through his magnifying lens as a "mother" aphid gave birth to tiny living progeny. Though others had seen this before (he cites Antoni Leeuwenhoek in addition to Geoffroy), he recounted his own experience as an object lesson for his readers who might wish to search out birthing aphids in their own gardens:

> I observed the largest aphids attentively with the loupe, and I did not have to observe them for long before perceiving several with a small greenish body at the anus or close to the anus, though the aphid itself was black. This little body was oblong, shaped something like a slightly flattened egg. I fixed my gaze on one of these small bodies, and I saw it emerge insensibly, further and further out of the backside of the aphid. It still resembled an egg. But finally when it was even farther out, when, to judge by its oval form, it seemed to have only the little end still inside the body of the mother, I realized that what I had thought until then to be an egg was actually a live insect, equipped with several legs. Its legs separated themselves little by little from the belly, along which they had been extended; I saw them move in different directions, perhaps in order to help pull the head out from inside the body of the mother, where it was still stuck.[8]

After witnessing many such births in different species of aphid—"once one has seen a fact of natural history, it is usually easy to see others like it again"—Réaumur carefully crushed some of the mother aphids and found a line of "fetuses" in various stages of development, reminiscent of eggs arrayed inside the bodies of hens.[9] As usual, he explained how readers could learn to see similar things for themselves: they would recognize aphids on the verge of reproducing by their larger size and the "stretched" or "inflated" look of their bodies. Some species did not produce offspring when confined to jars, but could be "forced to give birth" by the simple expedient of manually squeezing the live young out of the mother. "To succeed with these observations," he explained, "you must choose winged aphids of the largest species, such as those of the cardoon, those of the rose bush, etc. All the ones that I took from leaves, whether I pressed them gently or crushed them, showed me that they had bellies filled with young."[10] In most of the larger species, reproduction could be seen with the naked eye. "But if you want to use a magnifying lens, you will distinguish perfectly well the opening of the part where the insect emerges. It is shaped like a funnel with the flare on the outside; its border is white."[11] As observations proliferated of different kinds of aphids, all giving birth to live and fully formed insects, some fundamental questions remained. Both winged and wingless varieties—which some earlier observers had suggested might be of different sexes—produced young, so all aphids seemed to be female. "Are the two sexes united in them, as they are in snails? Even this does not suffice, it seems, [since] we see snails copulating. And no matter when I observed aphids, whether with or without wings, I never saw any copulation."[12] This seemed especially surprising, since (unlike bees, or other insects whose mating took place hidden from view) aphids lived in easy range of the naturalist's magnifying glass. Thus this painstaking work seemed to support the views of earlier observers who had categorized aphids as a special form of self-sufficient hermaphrodite. Though he did not consider the question of the aphid's sex to be settled definitively, Réaumur decided that if aphids do mate, it must happen before the mature stage, and this would be enough to make them exceptions to the general rule.

Although Réaumur designed an experiment to determine whether these insects could multiply without copulation, he was not able to bring it to a definitive conclusion. Nevertheless, he told the story in considerable detail "because others may be able to make the same trials with more success."[13] He planted a young cabbage shoot in a jar and placed a "mother aphid" taken from a mature plant on the new leaves. As soon as she gave birth to a new aphid, he removed her from the jar and left the newborn on its own. He covered the jar with gauze to allow air to enter while keeping other insects out. In several iterations of this experiment the young always died before reaching the age when they would normally reproduce. "If an aphid raised alone in this way did produce other aphids, it would be without copulation, or it would have had to copulate inside

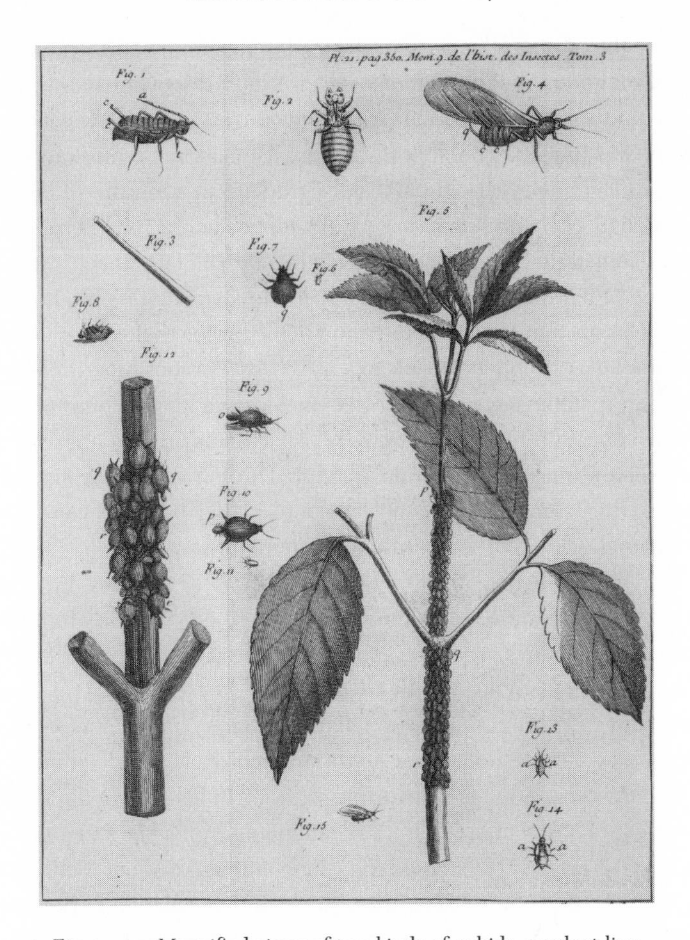

Figure 5.1. Magnified views of two kinds of aphids, or plant lice: newborn emerging from "mother" (figs. 9 and 10). Engraved by Philippe Simonneau from drawings by Hélène Dumoustier. Réaumur, *Mémoires pour servir à l'histoire des insectes*, vol. 3, plate 21. Courtesy of History and Special Collections Division, Louise M. Darling Biomedical Library, UCLA.

✲

the belly of its mother."[14] After four or five failed attempts, he gave up, though he assured his readers that it should be possible to devise a way to keep the aphids alive longer.

Réaumur's open invitation to repeat these experiments did not go unheeded. At least three assiduous observers picked up his suggestion. These three—Charles Bonnet, Abraham Trembley, and Pierre Lyonet—became in due course the most productive and engaged of Réaumur's correspondents, the source of innumerable observations, descriptions, discoveries, drawings, and novel techniques. Bonnet, to start with the youngest of this new generation of observers, was then studying law in Geneva. He had first written to Réaumur in 1738, an eighteen-year-old philosophy student, in a rapture of enthusiasm for the "delights" opened up to him by Réaumur's books: "A caterpillar is now for me an animal that cedes to no other in beauty; the facts that it presents me at the time of its transformation and during the course of its life hold for my mind the charms that I can no longer find in the ordinary pleasures of youth."[15] In this first letter, Bonnet sent his extensive observations of several common kinds of caterpillars as an act of homage, including a long account of a species he had first encountered in Réaumur's books. He was proving himself not only as an observer, but as an assiduous reader. The Parisian naturalist received many admiring, sometimes sycophantic, letters, but Bonnet's were the most prolix and meticulous of the lot. In a neat but minute hand, the first letter filled seven folio pages—brief compared to some of his later missives. From the beginning of this correspondence, Bonnet worked to establish his credibility as a naturalist in Réaumur's style, someone who noticed and recorded everything, no matter how small or apparently trivial.

Like many of Réaumur's correspondents, Bonnet could pursue natural history only in his leisure hours, but he did so vigorously, and kept voluminous notes on everything he saw. Réaumur recognized in his acolyte the kind of energy for minute observations that would undoubtedly lead him to notice novelties among the insects. He advised the young enthusiast about techniques and materials, like the solution of spirit of wine and sugar that worked best for preserving specimens. Soon Bonnet was casually referring to himself as one of the brethren of observers, sharing predilections and habits with his mentor.[16]

When he became an author in his own right some years later, Bonnet started off by recollecting that he had started out "trying to see again [*revoir*] after M. de Réaumur. I followed him, so to speak, step by step."[17] Bonnet admonished his own readers to replicate his results, just as he had done for those he had read in Réaumur. He had no more interest than his mentor in detailed taxonomy; he wanted to make his discoveries accessible and to inspire his readers to further refine and multiply them. "My main goal in publishing," Bonnet wrote, "was to give other people the opportunity to verify them and to push them further. I wish others to see after me, to correct me in all the places where I may have made a mistake. . . . I will even take great pleasure in providing them with all the clarifications they might need in order to repeat my observations more easily."[18] He was quite explicitly modeling himself on Réaumur's communication style, hoping to continue the cycle of emulation into another generation.

In the spring vacation of 1740, when he was staying at his family's country house outside Geneva, Bonnet started to work with aphids, to see if he could succeed where Réaumur had failed, by isolating a newly born insect and watching for it to reproduce. As soon as he saw an aphid emerge from the body of its mother, he moved it to a small branch devoid of insects; this he kept in a flowerpot of dirt, covered with an inverted glass vase to keep out passing insects (see fig. 5.2). He moved the jar into his study, where he could keep it under close surveillance far from any other aphids, hoping to keep it alive to maturity. Bonnet thought that Réaumur's solitary insect might have died due to lack of air circulation, so he took care to remove the cover from time to time. For a little over a month, he watched obsessively and recorded all the insect's movements in excruciating detail, eventually filling almost a hundred pages.[19] By his own account, he checked in with the aphid every hour when possible, and sometimes "several times in the same hour," from 5:00 in the morning until 10:00 in the evening. The little insect shed its skin four times, a maneuver he described exhaustively, and he watched it grow, walk, eat, excrete, and so on. "You would laugh, sir, if I told you what terrible distress my aphid caused me at its last change of skin. . . . It was so swollen and shiny that it looked to me like those aphids that are nourishing a worm inside their bodies. What contributed even more to make me believe this, and what increased my chagrin, was that it seemed to remain entirely motionless." All was well, however, and after the last molt, on the thirteenth day, "at 7:30 in the evening, my aphid [*puceron*] became a mother [*puceronne*] and from then until the 21st she produced ninety-five aphids, all living, and most of them born before my eyes."[20] Though this was not exactly unexpected, it was gratifying nevertheless to take a conjecture and turn it into a "demonstrated truth." Bonnet's excitement did not keep him from noting down the time each aphid emerged, removing the newborns so he could continue to keep track accurately as the numbers mounted. Eventually, he had to go into the city for a day, and when he returned early the next morning he was distressed to find that his aphid had wandered off. (He had left the jar uncovered, to prevent it from suffocating in his absence.)[21]

Bonnet's letter, with the detailed story of the prolific aphid, ran to twenty-three pages, including two pages of tables showing the date and time of each delivery (estimated, in the cases when he was not present at the moment of birth). Réaumur recognized his own role in inspiring these observations: "I could hardly be anything but satisfied that you took on the task of undertaking the observations that I did not have the time to do, and on which I invited observers to exercise their patience."[22] He had no doubt that Bonnet had witnessed an exception to "the law of copulation," though he also insisted that the observations would have to be repeated with as many kinds of aphid as possible. Bonnet was already seeking out other species; Réaumur also wrote immediately to Bazin, in Strasbourg, asking him to replicate the experiment as soon as possible. Bazin chose a species that matured quickly and was able to see offspring from his isolated insect in nine days. He also took the logical next step, removing one of the second-generation newborns and placing it in isolation to see if it would produce young of its own. But he had to leave home before this new aphid had matured, so he was not able to track successive generations of parthenogenetic births. The experiment was soon completed successfully, and independently, by Pierre Lyonet in The Hague.[23]

Bonnet, like Bazin, undertook his observations in the sequestered space of his study or his bedroom. He had to isolate himself along with his captive aphids in order to keep track of developments in his jars; indeed, when he returned to society his aphid escaped, bringing the episode to an end. The letters and reports of the naturalists are littered with similar endings, observations in progress interrupted by the obligations of daily life—a trip to town or an illness or a visit from a friend. The ideal of exclusive engagement with the insect subject was disrupted in other ways as well. Bonnet filled the pages of his notebook alone in his study, but he was soon reporting his victory to Réaumur and to plenty of others as well. With the turn from watching to communicating, the enclosed space of the study opened up into conversational and epistolary exchanges.

Bonnet's account of the parthenogenetic birth of ninety-five aphids served as his entrée into the Paris Academy of Sciences. Réaumur read from his letter to the assembled company, and shortly thereafter the academy named the Swiss student an official correspondent, in recognition of the significance of his discovery.[24] This was validation beyond Bonnet's fondest dreams, and certainly inspired him to continue along the same lines. He felt an obligation to continue his legal studies, but in the autumn vacation he devoted himself to aphids collected from oak trees, a species familiar to readers of Réaumur's third volume.[25] These insects were large enough to observe with relative ease, and very soon he had something novel to report: for the first time, he had witnessed aphids mating. The males were smaller than the females, with wings, and noticeably "ardent" in pursuit of females. He also noticed that not all of the newborns were fully formed; the mother sometimes emitted oblong "fetuses" that she deposited on a leaf next to each other, "just as butterflies deposit their eggs."[26] Bonnet was singularly reluctant to view these objects as eggs. After all, insects could be either viviparous or oviparous—but not both. Réaumur had explicitly laid this out as one of the "principles" of the natural history of insects:

"No species of insect is neither oviparous nor viviparous; every species perpetuates itself either by laying eggs or by giving birth to living young."[27]

Bonnet looked for status and recognition from Paris, but he also shared his discoveries locally with a circle of interested Genevans, including his philosophy and mathematics professors Gabriel Cramer and Jean-Louis Calandrini, and with his cousin Abraham Trembley, who was living in The Hague as tutor in the Bentinck household.[28] Geneva and The Hague soon became centers for the production of striking new results, new techniques and instruments, and eventually new books. Trembley, ten years Bonnet's senior, had been equally inspired by reading Réaumur, though he was not so brash about putting himself forward and first wrote to Paris only after his cousin prodded him to do so. Trembley had also taken note of the unresolved matter of the aphid's reproduction when he read Réaumur's third volume, shortly after it appeared, and had planned to explore it himself.[29] In the fall of 1740, he was gathering aphids for experiments he hoped to do on the effect of winter temperatures on these insects.[30] Trembley informed both Réaumur and Bonnet about the work of his friend and neighbor Pierre Lyonet, who was finally pulled into the web of correspondence somewhat later, in 1742. The letters exchanged by Réaumur, Lyonet, Trembley, and Bonnet provide a rich source for the strategies and techniques employed in natural history circles far from Paris. These letters mediated collaboration and competition and communication, feeding into the printed texts—which sometimes took passages verbatim from letters—that eventually defined the field for a wider public, sparking even more observations.

Over the years, Réaumur had numerous informants, assistants, friends, and acquaintances, many of whom witnessed or collaborated on his experiments, as we have seen. The letters of Swiss and Dutch insect enthusiasts show graphically how his books were taken as guides to new kinds of practice. Both Bonnet and Trembley referred to Réaumur as their "master" and themselves as "student" or "pupil"; in both cases an early and intense engagement first with the books and then with the naturalist himself led to their establishment as naturalists in their own right, and eventually as authors too. Lyonet, though not under the direct sway of Réaumur's authority when he started collecting and experimenting, also exemplified the Réaumurian ideal—wary of theoretical speculation, meticulous in observation, ingenious at devising experimental setups. He went far beyond Réaumur in his brilliance as an anatomist (of insects) and artist and engraver, all skills he honed in his leisure time, and the fruits of which he sent to Paris. Of these three men, all of whom shared their observations liberally with Réaumur, only Trembley ever met him face to face.

LYONET'S APHIDS

Trembley told Réaumur in one of his very first letters that Lyonet had witnessed the birth of aphids from a solitary individual in the previous summer, the same season when Bonnet was watching his first captive aphid.[31] Lyonet, a lawyer for the court in The Hague who studied drawing and portraiture in his free time, took up natural history as a kind of avocation. Sometime in the mid-1730s he began to observe, collect, and draw insects, tracking them through their metamorphoses whenever possible, gathering material for what he hoped would be the definitive guide to the insects of his region.[32] Like Bonnet and Trembley, Lyonet had been intrigued by the inconclusive experiments with aphids when he read Réaumur's third volume. The 1740 season happened to be a big year for aphids of all kinds. In September, unaware that Bonnet had already been keeping solitary aphids, Lyonet collected several different species and witnessed not only parthenogenesis but mating and egg-laying as well.[33] Trembley and Lyonet were discussing and comparing their observations in person; they often observed together. Lyonet was not yet directly in touch with Réaumur, and Bonnet learned of these experiments only through occasional reports from his cousin.[34] In fact, Trembley operated as a go-between, relaying results and messages between Lyonet, on the one hand, and Bonnet and Réaumur—who were also corresponding with each other. All of them were working on aphids at the same time, though not in a coordinated way.

While Bonnet composed his obsessively complete record for communication to Réaumur and the Paris Academy, Lyonet wrote his notes only for his own reference. They are rather more casual, and certainly less extensive, though they describe anatomy and behavior clearly and make particular reference to surprising or striking phenomena.[35] Lyonet was adept at keeping his captive aphids alive through several generations, and the question of fertilization was never far from his mind. Watching the offspring emerge hind end first, with the head remaining for some minutes inside the mother, he wondered whether

> perhaps by a very strange kind of copulation [*accouplement*] the mother communicates fertility to her offsping. To confirm this, I pulled one of the young gently from the body of the mother at the moment when it was held there only by the head. For all that, this little aphid was no less capable of producing offspring, although raised in isolation; it produced one before my eyes on the fifth of October.[36]

Where Bonnet had trained his gaze on a single reproducing insect, Lyonet kept several in isolation at once, following their asexual reproduction through multiple generations. In the fall, near the end of the season, "by chance," he found a colony of aphids on a twig in a ditch and noticed that some of them seemed to be mating. He then witnessed pairs mating in his captive population, which he could watch more closely, and confirmed with his magnifying lens the physical connection of the two individuals during copulation. "The winged aphid on top of the other held its back end curved backward and tightly joined to the back end of the other."[37] He was not, however, able to keep the insects alive to see what would happen in the following weeks.[38]

Both Lyonet and Trembley saw what Bonnet called "fetuses" deposited by their aphids at the end of the season. Lyonet recorded his surprise at finding what he thought must be eggs in the jar where he kept an aphid taken from a cardoon plant. He had seen two live offspring emerge; several days later, he found in the jar not young aphids but eggs—one on the leaf and several on the glass. "This surprised me, . . . I had difficulty persuading myself of it since I thought, based on Réaumur, that all aphids are viviparous."[39] To follow up, he put a dozen of the aphids in a jar and soon found fifty eggs on the glass, and not a single newborn. At this point he could not help wondering whether aphids behaved differently as winter approached:

> This made me think that perhaps the last of the generations of aphids in the year produces eggs in place of young, eggs that last the whole winter, and what would support this idea is that I do not understand how newborn aphids could survive the rigor of the seasons out in the open; these new aphids are too delicate to go underground, and bad weather and cold kill them.[40]

Lyonet showed the aphid eggs to Trembley, who then spotted some in his own jars. Lyonet did not worry about communicating his results at this point; he let Trembley inform both Réaumur and Bonnet of the progress of the Dutch observations."The fetuses mentioned by Bonnet are also known to us," Trembley told Réaumur. "M. Lyonet showed me some in October, produced by a cardoon aphid that had also given birth to live offspring."[41] Trembley also told his cousin that Lyonet had seen the insects mating and invited Bonnet to join them in repeating all of these experiments in the spring.[42]

When the Dutch publisher Pierre Paupie commissioned a French translation of Lesser's *Insectotheologie*, he asked Lyonet to review the text for factual errors and agreed to print extensive notes and comments (see chapter 1).[43] Lyonet's only mention in print of the unexpected phenomena he had observed in aphids was buried in one of his notes to Lesser's book, in a discussion of the universality of sexual reproduction. "However general this rule seems to be," he noted, "we are not yet completely certain of its universality."[44]

Leeuwenhoek, Cestoni, and Réaumur had all looked for evidence of two sexes in aphids to no avail; none of them had seen anything other than the adult "mothers," nor had they seen the insects mating. Lyonet's own experiments with mature aphids confirmed their asexual reproduction, but he also mentioned that he had seen aphids mating as the season progressed toward winter. His comment reads very much like the kind of narrative observation report familiar from Réaumur:

> Having pushed my experiments up to the time when the leaves begin to fall, and no longer doubting the truth of the thing [i.e., asexual reproduction], I was suddenly undeceived when I least expected it. I had gathered together all the aphids that my solitary aphids had produced for me, and I had established a little colony on a bit of a willow branch that I kept fresh in a glass of water . . .[45]

Réaumur knew about these observations from Trembley and asked him to encourage Lyonet to write to Paris directly. The two exchanged their first letters in early 1742, after the notes and illustrations for *Théologie des insectes* were completed but before it was published. Réaumur was finishing his sixth volume and hoped to include Lyonet's surprising observations in a new chapter on aphids that would bring readers up to date on all that had happened since the inconclusive experiments four years earlier. Lyonet sent the text of his note, along with a sheet of drawings, some of which were engraved for plate 47 of Réaumur's volume (see fig. 5.2). At this time, both Bonnet and Réaumur were still confused about whether the aphids laid eggs in the fall; Réaumur had seen oblong objects that he called "aborted fetuses," and Bonnet adopted this term.[46] For his part, Lyonet had found eggs laid by some of his aphids, and even recalled that some had hatched in the spring:

> Since I am familiar with more than one kind of viviparous aphid in which the last generation of the year lays eggs that last through the winter and hatch in the spring, I would have very much wished to examine whether the aphids in question, after having mated, instead of producing live young also laid eggs, as there is some reason to presume. But I have not been able to verify this, because my aphids wanted to leave their branches after mating, and drowned in the glasses of water.[47]

Réaumur could not help but be impressed by the "extremely singular" things that Lyonet had managed to see. They seemed irrefutable proof that aphids were not like other insects, that they might even be called a new kind of "marvel." It seemed amazing enough that they could do without mating. "But is it not an even greater marvel," he wrote to Lyonet, "that the season's final

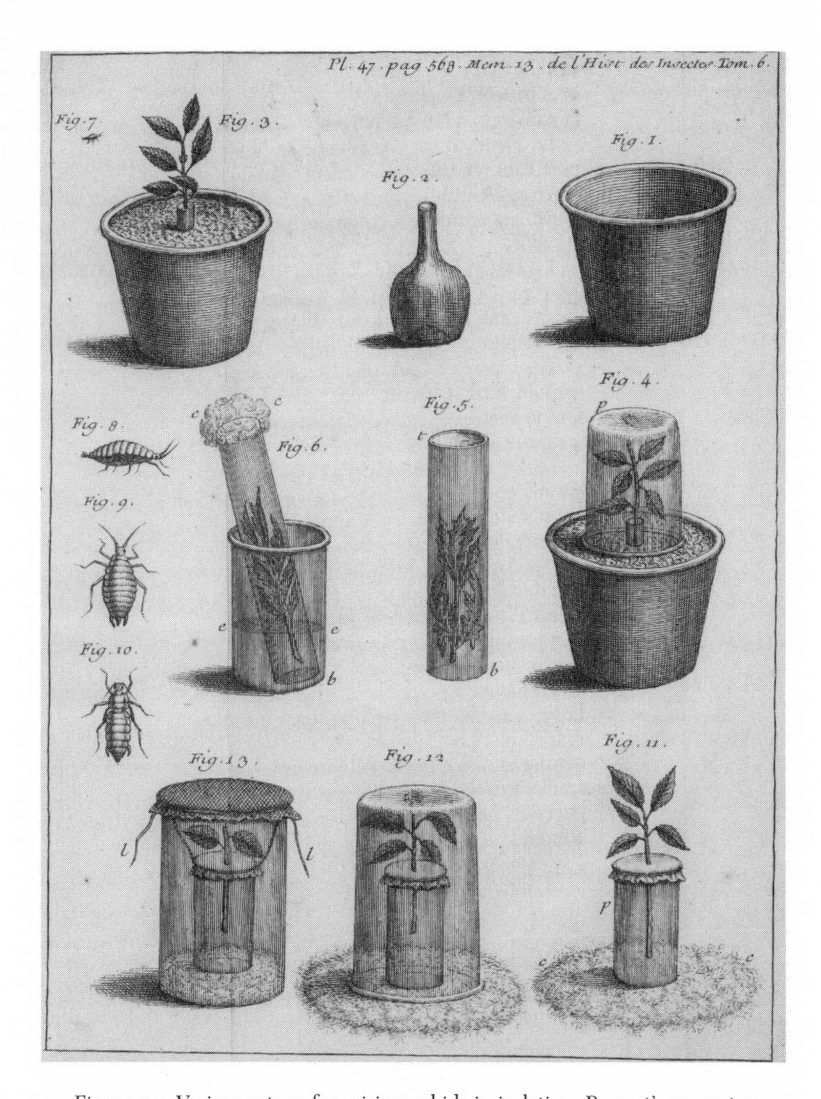

Figure 5.2. Various setups for raising aphids in isolation: Bonnet's apparatus (figs. 1–4); Trembley's arrangement (figs. 5 and 6). Figs. 7–10, engraved from drawings sent to Paris by Pierre Lyonet, show wingless aphids he had observed mating on a willow tree. Réaumur, *Mémoires pour servir à l'histoire des insectes*, vol. 6, plate 47. Courtesy of History and Special Collections Division, Louise M. Darling Biomedical Library, UCLA.

generation of animals copulate, when they were preceded by several others for whom all mating was unknown?" There remained the question of whether these couplings were necessary to the aphids' survival, and if so, in what way. "We should no longer be surprised to find that in the same type of insects there are some viviparous and some oviparous species. The two-winged flies provide us with examples. But it does seem very surprising that within the same species some individuals are viviparous and others are oviparous, according to the season when they work to perpetuate their species."[48] Since neither he nor Bonnet had ever seen aphid eggs hatch, and indeed did not think these "oblong bodies" were eggs at all, Réaumur asked specifically what Lyonet had seen in this regard.

Lyonet answered tentatively:

> Mating must certainly be necessary for them, since they do mate; but the effects will perhaps not be easy to discover. One could conjecture that it serves to render fertile the offspring of one whole year and converts the viviparous aphids into ovipares in order to preserve the species during the winter by means of eggs. But these are still no more than conjectures, and we know how often conjectures can deceive us.[49]

Lyonet's observation notes show that he had been thinking along these lines for some time, and the idea that a viviparous insect could become oviparous before the winter does not seem to have troubled him unduly. He recalled having seen in the spring, in the jars where he had kept some aphid eggs over the winter, young insects evidently hatched from the eggs—although they were dead when he noticed them. He had not recorded this in his notes, however, and worried about whether he was remembering correctly. "For greater certainty, I will repeat the experiment," he promised.[50] But that summer he was too busy with his professional obligations to do much with aphids, and he dropped this line of investigation.

In the meantime, Réaumur finished his updated essay on aphids and inserted it into his sixth volume. He recounted everything from the reports sent by Bonnet, Trembley, Bazin, and Lyonet, with special emphasis on Bonnet's first observations in the summer of 1740 and their enthusiastic reception at the academy. Réaumur gave the story a nice narrative arc, with the kind of circumstantial detail his readers had come to expect, assigning credit to various observers for different pieces of the puzzle. He even printed Bonnet's two pages of tables, graphic evidence of the young man's perseverance and meticulous attention.[51] The reader learned about several alternative arrangements for the containers, branch, water, and insect to keep the solitary aphid alive while keeping it visible. The surprising fact of virgin birth had been witnessed by many eyes: "Finally they all did experiments that had the desired result: they saw and saw again aphids of different species giving birth to [their] young without having had contact with any insect of their species from the time of their own birth, and even without ever having been near any."[52]

Even though Réaumur had plenty of eminently trustworthy evidence from his correspondents, "I would nevertheless deserve reproach if I had not sought to assure myself with my own eyes of a truth demonstrated by experiments for which others had taken responsibility at my instigation."[53] Once again, he chronicled his failures as well as his successes, encouraging readers not to give up if their first efforts came to nought. His first attempts failed because he was not able to check on the insects frequently enough—too much moisture gathered on the inside of the glass containers, and the aphids died. He varied the setup to allow better ventilation, noting that this would be the best method for "all those who must engage in regular duties every day for several hours in a row"—in other words, everyone without the leisure to watch an insect uninterrupted, around the clock.[54] "I had occasion to be happy with a poppy aphid, born while I watched on the 12th of June around noon, and which I enclosed immediately, with all the precautions explained above: . . . Finally, aged less than seven days, . . . it gave birth in front of me to a living offspring in good health."[55]

In print, Réaumur presented the story of the aphids as an exemplary case of the deployment of observational resources, with multiple naturalists in different places exerting their attention and ingenuity to seeing the previously unseen. The successful series of corroborations also showed off the role of Réaumur's book as a catalyst for the production of new, and newsworthy, knowledge. After all, reading the original account had inspired Bonnet to give his unwavering attention to the solitary aphid in the first place. Following up on the successful work of others, Réaumur played multiple roles at once: arbiter of knowledge in Paris and author of the first account in print, but also reader and facilitator of the other reports. Even with the new observations of parthenogenesis, the subject remained open-ended in certain respects; when his book went to press, no one had yet seen aphids being born from the eggs.

Réaumur referred to fertilization in aphids as one of the most peculiar facts of natural history, "a singularity interesting for natural philosophers [*physiciens*], and even for metaphysicians, and well suited to justify the time spent on observing the tiniest insects."[56] Above all, this extensive investigation, expanded to many different species by multiple accomplished naturalists working over three annual cycles, showed the risks of jumping to conclusions about the generality of "laws" of nature. The most commonly accepted of these laws was the necessity of the congress of two individuals for the production of offspring. Even in hermaphroditic species like snails, two parents still had to copulate—individual snails or worms cannot fertilize themselves. Another law of the insect world, familiar to readers of Swammerdam, held that fertilization of female insects happens only after the final transformation to the adult form. But the new results showed unequivocally that aphids could be sequestered before their final molt and still produce young; however they acquired their fertility, it did not happen at the adult stage. Réaumur, resisting the idea that the "aborted fetuses" observed by himself and Bonnet might actually be eggs, was reluctant to accept Lyonet's suggestion about a cycle of viviparity, fertilization, and oviparity timed to the seasons. "What purpose does copulation serve?" Réaumur wondered. "Is it to replenish the exhausted fertility in these insects who from mother to mother had been virgins during several generations?"[57] Lyonet had not been able to definitively determine the effect of copulation—his captive aphids died after they had mated, without producing any young. Would the products of parthenogenesis differ in some way from those born from insects after mating?

Réaumur cited both Bonnet and Lyonet as having seen "little oblong bodies" or "a sort of egg," though he doubted if these bodies were eggs at all.[58] Recapitulating his exchange with Lyonet, Réaumur commented that if they were to hatch, which had not yet been observed, the naturalists would have yet another peculiarity on their hands, with the sequential viviparity and oviparity of the same species, according to the season. "Therefore even though these little insects have been well studied, they deserve to be studied even more."[59] Lyonet went on to do just that and eventually witnessed aphid eggs in the process of hatching. He told Réaumur every circumstantial detail leading up to this crucial observation, with advice about how to find the eggs. A trail of ants had led him to a crevice in the bark of an oak tree, where he found "bodies exactly like those I called aphid eggs, and they had the same disagreeable odor of the aphids of that tree." He took a dozen eggs home, keeping them moist on a piece of cork set in a saucer of water, covered by a glass. "Four of these eggs shriveled up, the others stayed in good shape, and from one of them, last Sunday the 7th of this month at noon," an aphid emerged:

> It was only half out of the egg when I noticed it, and it took more than another half hour for it to disengage itself entirely. I asked M. Trembley to come see it, we examined it with a loupe, and we found it to be just like the large species of oak aphids with the long proboscis, that you describe near the end of the *mémoire* [on aphids] in your third volume.

Since it was a species familiar to Réaumur, Lyonet had no doubt that he would be able to find the eggs in oak trees where he had collected aphids the year before.[60]

TREMBLEY'S POLYPS

When Bonnet encouraged his cousin to do as he had done and correspond directly with Réaumur, Trembley had already taken up natural history. For some time, he had been keeping live insects, including aquatic species, in jars and boxes in his study, intent on seeing their transformations and maneuvers. Living at the country estate of the wealthy Bentinck family as tutor to the children, he modestly described himself to Bonnet as no more than a "schoolboy" with regard to natural history, "undertaking to follow Réaumur, whom I have often had occasion to admire."[61] Trembley found the insect world a constant source of pleasure, as Réaumur intended his readers to do, and found this kind of study "very appropriate to forming the mind." What better occupation for a tutor and his young charges?

Réaumur could hardly have predicted that his new correspondent, who had been gathering and watching all the caterpillars, flies, and aquatic insects he could find in the Dutch countryside, would soon be reporting something even more unexpected than the life cycle of aphids. Much as Bonnet had done, Trembley described some of his observations in his introductory letter as a way of demonstrating his seriousness and his skill. He sent specimens of a wool-eating caterpillar, with its moth, and told Réaumur about the aphids that he and Lyonet had been following. In his second letter, a few months later, Trembley detailed his replication of Bonnet's aphid observations, using different species, and correlated the frequency of births with ambient temperature. He outlined some experiments he planned for the coming winter on the effects of cold on insects, and enclosed some eggs from a nocturnal butterfly. "Before closing," he went on, "I cannot refrain from taking a moment to tell you about the object that occupies me most at present."[62] This was a small "organized body" that he had found the previous summer, though he could not identify it. Indeed, he could not even decide whether it was animal or plant. He had noticed several of them in his glass containers of stream water, where he was keeping various aquatic plants and insects, and had been watching and experimenting with them ever since.[63] The mysterious "beings" had small, green tubular bodies with fine threads like tentacles, or perhaps legs, at one end, less than half an inch long overall. They attached themselves to the leaves of the water plants, but also to the bottom and sides of the glass jars, where Trembley could watch them contracting and expanding and moving in various ways. He remarked the odd fact that they migrated toward sunlight. When he sliced halfway through one of the bodies, the cut grew back together in a few days. But he had only recently discovered "the most remarkable fact" after cutting one of them into two pieces: both parts remained alive, and each grew back its missing half over the course of a few weeks.[64]

At first, witnessing this regeneration, Trembley thought of the way crayfish could grow back severed legs and antennae, alluding to a phenomenon Réaumur had studied in 1712.[65] But in this case, rather than regrowing a missing appendage, each of the two halves grew back into a whole. This suggested the way some plants can reproduce from cuttings, but in other respects these aquatic "beings" clearly behaved like animals. Trembley's excitement was palpable, even as he wondered if these little bodies were already known to other observers. "I still have a great number of experiments to do. I will cut some of them into several parts. I will gather several of these parts to see if they seek the daylight as a whole animal does. I do not know what to call it. I still do not know everything there is to know about the way it multiplies."[66]

Over the next few months, Trembley continued his experiments with regeneration and explored every aspect of the anatomy and physiology of these "enigmas." He reported a litany of unexpected and inexplicable results—he obtained three fully functioning individuals by cutting one body into three pieces; the interior and exterior membrane of the body's tube enclosed a green gel-like substance; the body contracted and expanded in length, causing a change in color as well as shape; the flared end near the tentacles seemed to be a receptacle, sometimes empty and sometimes full; they moved along the glass surface in the manner of inchworms. The swaying motion of the tentacles suggested sensitive plants, but the locomotion made him think they must be animals. Trembley communicated to both Réaumur and Bonnet his sense of never knowing what he would find when he checked his jars. He mentioned that Lyonet, who kept aquatic plants and animals in tubs full of water from the same location, had seen the unfamiliar bodies as well, but he had assumed they were plants and had not experimented further until Trembley alerted him to their peculiarities. In fact, the two men often collected and observed together, testing various techniques for viewing the little bodies with lenses mounted on flexible arms.[67]

Trembley's captive population stayed alive over the winter, and in late February he noticed what turned out to be the beginning of the reproductive process. He informed Réaumur a few weeks later:

> On the 25th of February, in the morning, I saw a little body that climbed up one side of the large glass where I keep my supply. I noticed approximately

> in the middle of the body, on the exterior skin, a little protuberance of a deep green. . . . I observed my little body very often between 10 in the morning and 4 in the evening. . . . So that this body would not become lost among the others, I pulled it out of my large glass and put it by itself in a bottle of a very transparent glass. I placed it so that it attached itself to one of the sides, and in this way I could watch it easily by daylight and by candlelight.[68]

The appearance of the green lump in the skin, and the shape it took on, made him wonder if it might not be a nascent "organized body." Over the next few days, with mounting excitement, he watched the lump lengthen out into a cylindrical shape. On the fourth day, "I found it even more elongated. It was about half a line in length, and the same color as the body. Finally at 10 in the evening, I saw what I was waiting for with the greatest impatience, that is to say the threads that started to emerge [from the end]. You can well imagine, sir, that it was not without pleasure."[69] When he wrote, Trembley was still waiting for the secondary body, with its newly formed threadlike tentacles, to separate from the original one. "They are connected by no more than a point. I will not neglect to put my newborn in a jar by itself."[70] This budding process made him think that these things must be plants after all.

Needless to say, Réaumur was interested in this "plant animal" or "animal plant" and asked immediately that samples be sent by post. After waiting for the weather to warm up, Trembley sent about fifty in a sealed bottle, but they were dead on arrival. Before trying again, he punched air holes in the cork and took a jar of them for a ride into the countryside, "at a trot," to simulate the jostling of the coach. With this new method of packing, they survived the strenuous journey to Paris.[71] In the meantime, Réaumur had read Trembley's observations of regeneration and budding to the Academy of Sciences.[72] As soon as he received the bottle, and realized that the specimens were still alive, he spent the evening watching them by candlelight, and found that they behaved exactly as Trembley had reported. Only after seeing them move around did he pronounce them to be animals, reversing his earlier assessment, and quickly came up with a name for them. He called them polyps, a designation Trembley gladly accepted, and took them along with him to show "in person" to the academy.[73] At this point, Réaumur had not yet tried the sectioning experiment:

> Although all those you sent me just now seem to be doing very well, I wanted to let them recover a bit from the fatigue of the journey, before making them submit to a cruel operation. I do not think, however, that the day will go by without one at least being cut in two. I am very impatient to see this prodigy with my own eyes, and I defer reasoning about it until I have witnessed it.[74]

It was not long before he had seen the "prodigy" himself; the next step was to demonstrate the polyp's regenerative capacity to the academy, and then to anyone else who wanted to see it.[75]

In Holland, Trembley was also demonstrating the striking peculiarities of regeneration and budding—to Lyonet, of course, and to various other friends and aquaintances. His employer, Count William Bentinck, was part of this circle, as was his old physics professor from Leiden, Jean Allamand. The peculiar phenomena in his jars were soon being discussed throughout the Dutch natural philosophical community.[76] As the witnesses to the polyp's strange abilities multiplied, Trembley continued to experiment intensively, finding new kinds of polyps and a seemingly endless string of novel and weird facts. He kept a large population in what he called his "nursery," in containers of all sizes, as he found new species. He repeated everything many times, and got more and more inventive and adventurous as time went on. One of the new species was large enough to cut into many pieces; new individuals formed from each of the pieces. Then he tried cutting them longitudinally, and watched as the edges curled together and sealed, forming two viable individuals from one. Eventually he was able to affirm that the tentacle end housed the animal's mouth, only visible when it was open to receive food. "It is very difficult to see the hole. I have seen it. But I have seen even more than that: I have seen polyps eating. They are carnivorous and very avid. They swallow little eels longer than themselves. One can see them curled up inside the body." He watched a mother and an attached offspring share a little eel, which passed from one body to the other. Once he had observed digestion, he used the digestion of eels as test of whether regenerated bodies are truly complete polyps.[77]

In one of his most striking experiments, Trembley turned the body of a polyp inside out, to see whether it would still be able to eat and reproduce. Once he had inverted the tubular animals, he strung them up with hog bristles, to keep them from turning back again. "Their body is a sack, which can be compared to a gut," he told Bonnet. "I turned it inside out as one turns a stocking or a glove. The inside became the outside, and the outside the inside, and the polyps on which I performed this operation live, eat, and reproduce."[78] Bonnet complained that this rather cryptic announcement of yet another "prodigy" could hardly satisfy Trembley's Genevan audience. "You never speak to us except in enigmas. Mr. Cramer, and all the savants and gentlemen who admire you here, make the same reproach."[79]

Bonnet had still not even seen a polyp for himself, two years after Trembley's initial observations, having failed to find them anywhere in the vicinity of Geneva. While searching the local ditches and ponds, however, he had found several species of aquatic worms that regenerated in a similar fashion when cut, and he observed them assiduously. Trembley was busy writing his long-awaited book on the polyps and asked Bonnet, "and all the *curieux* of Geneva," to wait for a full explanation until they could read the book and see the illustrations, then being made by none other than Lyonet. In Holland, Trembley was in the habit of supplying the creatures to anyone who wanted them; "and I indicate to them all the procedures I use to do my experiments, so that they may see them again [*revoir*] and improve on them." He had even taught Allamand to turn the polyps.[80] This exchange, just a fragment excerpted from a long and detailed correspondence, shows the widening circles of interest, discussion, and participation in the striking new phenomena being elaborated in and around The Hague and Geneva, as well as in France and England. The stream of surprising results from the manipulations of polyps and worms set off speculation and metaphysical discussion, especially in Geneva, where some of the observers, like Bonnet and Cramer, were especially prone to speculation. Some people worried about the implications of regeneration for the notion of an animal soul; others wondered if this would lead to materialism. Marc Ratcliff has demonstrated the crucial role of Trembley's policy of distributing polyps in the development of microscopical research across Europe; in Geneva, the phenomena he reported also fueled more abstract debates.

The burgeoning investigations of aphids and polyps fit perfectly into the kind of natural history practiced and then narrated by Réaumur. He crafted his books, with their combination of description (to allow identification), illustration, first-person narrative, and reports from other observers, both as record of the current state of knowledge and as inspiration or exemplar for his readers to follow. From their reading of these books, Bonnet, Trembley, and Lyonet all gleaned not just general inspiration and methodological precepts, but also specific questions and directions to guide their own experimentation. In their letters, the same dynamic worked at a faster pace, as they sent queries, results, specimens, and advice back and forth through the post and in the carriages of traveling acquaintances. Among some pairs or groups of naturalists, exchanges took place in real time, through joint observations and conversation. One element of this kind of natural history involved casting a wide net in the search for anything of interest. A naturalist had to take advantage of luck and contingent circumstances and be prepared to home in on whatever might cross his path—whether spider, mite, aphid, flea, or worm. Trembley noticed his first polyp when he was looking at other things in pond water; Lyonet discovered the regenerative capacity of water worms when he was feeding them to his dragonfly larvae.[81] Ideally, a naturalist working in this vein would have many investigations under way simultaneously. The relentless accumulation of empirical details did not preclude the bigger picture, however. Even those naturalists closely focused on the behavior of barely visible aphids occasionally found themselves addressing persistent problems like organic generation, metamorphosis, or regeneration.

REGENERATION AT THE SEASHORE

Trembley had no interest in rushing into print with his discovery and continued to experiment with different species of freshwater polyps for years before he was ready to publish his own book. He studied their digestion and feeding habits, modes of locomotion, and reproduction (by budding), with the goal of producing "complete histories." In the interim, he was happy to have regeneration introduced to the public by Réaumur in Paris and later by Martin Folkes in London.[82] Indeed, he hoped that more and more species would come under careful observation: "The land of polyps is so vast," he wrote to his mentor, "and I wish so heartily that it should be known, that I wish that hundreds of people were already occupied with traversing it. I hope that the preface to your sixth volume will have engaged several naturalists to devote themselves to finding and observing polyps, and that in a few years we will know even more marvels than we do at present."[83]

When Réaumur first showed Trembley's polyps to his colleagues in Paris, the botanist Bernard de Jussieu realized that he had seen a similar aquatic creature, though he did not know about its regenerative properties. These "reddish" polyps were the first of many related species found and sectioned by observers all over Europe in the next few years. In the summer and fall of 1741, naturalists and their friends were looking for regeneration everywhere. Bonnet turned to water worms when he failed to find polyps in Genevan ditches; Trembley and Lyonet also worked with various kinds of worms; Jussieu went to the Normandy coast looking for sea stars and other saltwater creatures that might regenerate when cut; and many of Réaumur's other correspondents searched for polyps in their neighborhoods, often settling for worms instead.[84]

Jean-Etienne Guettard, a newly minted doctor who had studied botany with Jussieu, had been lodging with Réaumur in his growing establishment on rue de la Roquette since shortly after the household's move there in 1740. Guettard did a lot of collecting for Réaumur's tubes and jars, and journeyed to Poitou with the household in September 1741 so he could work on the Atlantic coast near La Rochelle.[85] He was dispatched to the shore to investigate regeneration in sea anemones and sea stars, creatures that were often found with one or two legs missing. He soon reported that "your suspicions are confirmed here"; he had sectioned a variety of sea creatures in every imaginable way, much as Trembley had done with his polyps. Once he discovered that the amputations did not kill the animals, Guettard went at the task with a will:

> I began to cut mercilessly and cut a large number that I put in this tide pool. There were several sea anemones [*culs de chevaux*] around; I cut them. . . . I found another little tide pool with sea stars in it; I cut them in two. I found some sea anemones; I cut them. Finally, like victorious soldiers who after defeating their enemies chase the hussars, severing the arms and legs of those they encounter, I cleaved in two all I found, I strewed here and there arms, half-cut bodies, bodies cut into two pieces. In conclusion, sir, I will see whether, as fortunate as Perseus who found men after he sowed dragons' teeth, I will have or could hope to have complete bodies from those I cut in two.[86]

He found it difficult to keep track of all the severed parts and had to experiment with different ways of keeping them in salt water; some he brought to his lodgings, and some he kept in casks in the tidal area or in deeper water anchored to a seawall.

Guettard worked at the coast for fifteen days, not quite long enough to follow the complete restoration of the limbs he had amputated. He brought back with him some specimens in various stages of regrowth, but they could not be kept alive for more than a few weeks without a source of seawater to replenish the tanks. Réaumur asked for help from his old friend Charles-René Girard de Villars, the doctor who had observed with him in the 1730s in Poitou, and who had since moved his medical practice to La Rochelle. Villars agreed to collect, section, and keep track of sea anemones, marine worms, and sea urchins. He and Réaumur seem to have worked together at least for a few days; Villars refers to "our aquaria" and "our experiments." By mid-November, after Réaumur had returned to Paris, his collaborator still had live animals in eleven of his original thirty jars of seawater:

> Experience taught me that brown sea anemones are almost like amphibians in that they do not want to be covered by water all the time, and they tried to climb up the side of the container when it was too full. Other whole ones that I have in earthenware plates and to which I gave some pieces of algae that they seemed to like, are doing well and also move back when they have too much water.[87]

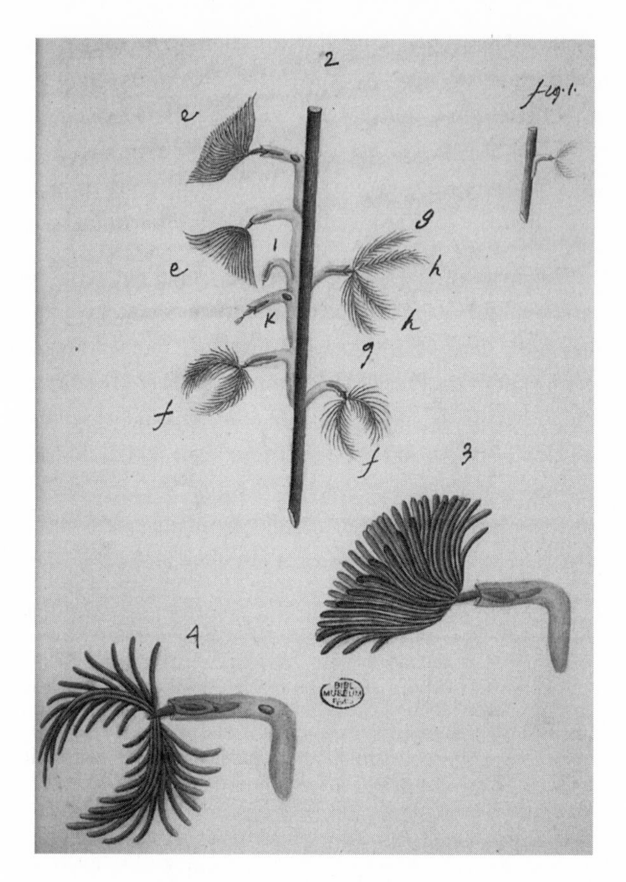

Figure 5.3. Hélène Dumoustier, plumed polyp observed at Réaumur. Ms. 972, BCMHN. By permission of Bibliothèque centrale du Muséum nationale d'histoire naturelle, Paris.

✲

He also tried a larger saltwater aquarium, a container with sand, mud, and rocks where he experimented with keeping many species. He had help from his friend and neighbor, Louis Richard, another local notable with a penchant for natural history. They planned to establish an aquarium in the tidal area after the worst of the winter cold had passed. "I will make all these outings with pleasure . . . I hope that this new year will bring a nice series of experiments."[88]

In the spring, Villars made excursions to different spots on the coast and gathered a new crop of creatures to be sectioned and kept in the saltwater tanks at the home of another friend.[89] After several months, he reported on the progress of the cuts:

> Of the great number of sea anemones that I had divided in different ways, many died; but there are still enough of them to show that their wounds healed well. Of all those that I sutured there are only two left that we can count on. The sea stars have not until now benefited from joining by suture; I have only seen a few on which one arm, or branch, has grown a little.[90]

Evidently, he had initiated a completely new set of experiments, to see if the arms would regenerate when the cut was sutured. In the fall of 1742, Guettard, Réaumur, and Hélène Dumoustier went together to the coast and met up with Villars.[91]

Inspired by the tiny freshwater polyp to look for regeneration in larger aquatic creatures, this motley group of observers, working sometimes together and sometimes separately, were showing "that such an inconceivable property was given to many species of animal."[92] Their letters, hinting at the elaborate observations and experiments performed around the tide pools and beaches of the Atlantic coast, give a sense of how this kind of investigation spread and developed, in some cases (as for Villars) altering the habits and rhythms of daily life. The vicinity of La Rochelle, like The Hague or Geneva, became a center of activity, with a variety of people clambering around on rocks and cutting off the arms of sea stars, or putting animals into tanks of seawater to be observed with magnifying lenses, or devising containers that could be washed by the tide.

POLYPS IN PRINT

News of the regenerative capabilities of polyps and worms spread by word of mouth and by letter, but it was only with the publication of Réaumur's sixth volume on insects, in 1742, that a larger reading public found an account, albeit a brief one, in print. At this point, Trembley was still working on his observations, holding off on publishing his book, and Bonnet was not yet ready to publish his experiments with worms. Lyonet was making the delicate drawings of polyps that he planned to engrave for the plates in Trembley's book, but he was in the early stages of this enormous task. In the meantime, it seemed a pressing matter to publicize the discovery, to answer "the large number of questions that I have been asked, whether verbally or in writing, about its reality." Réaumur introduced the little animals in his preface "to attest to the truth of a fact that engaged the curiosity of everyone who heard about it." He enumerated the many versions of regeneration in different species as a way of making the polyp look normal, and also in a move to claim this "inconceivable" property for the kind of systematic observation familiar to his readers from previous volumes, and to inspire at least some of them to go looking for polyps themselves.

The polyp, in the year since it had first been witnessed by the Academy of Sciences, had become a hot topic "at court and in town [*la cour et la ville*]." And for all the conversation, "I have not seen anyone who believed it on first hearing about it."[93] Réaumur was nearing sixty at this time; he never completed the remaining volumes of his great work, and this turned out to be the only one of his books to include the contributions of his prolific younger correspondents, Bonnet, Lyonet, and Trembley. In the case of the parthenogenesis of aphids, Réaumur had made the initial observations himself; with the polyp, he acted more as an impresario for the next generation of naturalists—though, of course, he had confirmed their observations himself. Réaumur's narrative of the polyp's discovery and his overview of its salient features anticipated the authoritative account by Trembley, published two years later. By putting it into the context of new work on regeneration in other species, as well as the updated account of aphids that owed so much to Bonnet and Lyonet, Réaumur was asserting his status as mentor to these newcomers while also indicating the direction of future research.

Réaumur wrote up the discovery and confirmation of the polyp's peculiarities as yet another exemplary case of the way natural history should be pursued. Trembley had come across the creatures by chance, but it was only because he knew how to look that he was able to take advantage of that chance. By implication, it was because he had been reading Réaumur that he knew how to look. He had been wary about jumping to conclusions, hesitating about whether the polyp was plant or animal, and unwilling to believe his eyes without multiple iterations of a given observation. Réaumur embedded the description of what Trembley saw in the story of the arrival of the polyps in Paris, and his own first sight of their regeneration: "I admit that when I saw for the first time two polyps forming themselves little by little from the one I had cut in two, I had difficulty believing my eyes; and it is a fact that I am not at all habituated to seeing, even after having seen it over and over, hundreds of times."[94] However surprising, Trembley had shown that the polyp's peculiarities could be produced over and over, and furthermore, that many other animals, once naturalists knew to look for them, exhibited the same property.

The polyp appeared in print again very soon, in Lyonet's notes to *Théologie des insectes*, published shortly after Réaumur's volume. Without challenging Trembley's priority, the Dutch naturalist was advertising his own role in the unfolding discovery, in the oblique form of his commentary on Lesser. In a note on insects that resemble plants, Lyonet referred slyly, and rather disingenu-

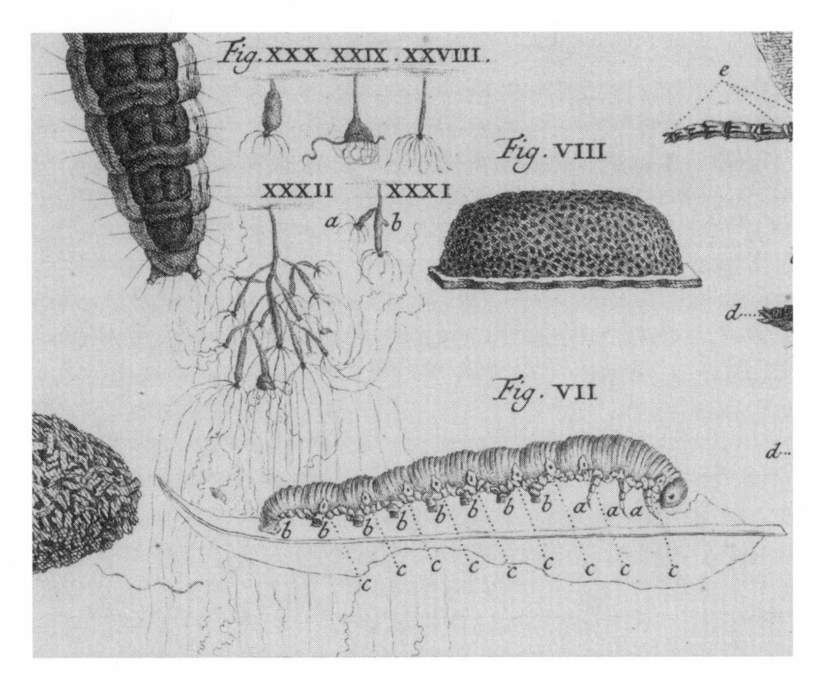

Figure 5.4. Pierre Lyonet, freshwater polyps (engraved by J. van der Schley). Detail of plate (see fig. 1.3) for F. C. Lesser, *Théologie des insectes* (1742). Ms. Fr. 99, Houghton Library, Harvard University.

✲

ously, to "that animal common in our ditches" as an example of something that looked like a plant but turned out to be an animal. Trembley, "a very attentive observer, who verified before me all the facts I have just reported," had taken months to decide whether it was a plant or an animal.[95] The reader found here, compressed into a few paragraphs, the essential points of interest about the polyp's behavior, structure, and reproduction by budding, but not a word about its ability to regenerate after sectioning. Lyonet never used the creature's new name either; but he did insert five tiny, exquisite images of it into the foldout plate illustrating different types of insect transformations. These images of the polyp, almost spectral compared to the many other insects in the plate, laid claim not to the original discovery but to the ability to make it visible. They subvert the whole notion of a taxonomy of transformations (adapted from Swammerdam), since they don't fit into any existing category. With the fine tendrils of their "arms" spilling down behind one of the caterpillars in the image, the delicate polyps could be read as teasers for the more elaborate illustrations

Lyonet was then making for Trembley's book, the definitive and complete history of polyps.

EXPANDING THE PUBLIC FOR NATURAL HISTORY

Trembley's book was published in Leiden in 1744, in an elegant quarto edition with thirteen finely engraved plates. Lyonet had drawn all the images and engraved most of the plates as well.[96] Trembley made a point of including tips for how to find, keep, and observe polyps. He wanted readers to go out and collect animal life from ponds and ditches, to see for themselves the marvels he was recounting. One of his eager readers was Gilles Bazin, in Strasbourg. Rather late in life, after years of observing and experimenting on insects, Bazin wrote a series of books designed for readers looking for less density of detail than they would find in the tomes of Réaumur or Trembley. He started with bees, transforming Réaumur's material into conversations between a didactically minded naturalist and his intelligent female friend.[97] Soon after Trembley published his *Mémoires* on the polyp, Bazin decided to write another of his "conversations" on the same subject. He published it initially as a pamphlet in Strasbourg and included it, a few years later, as part of a four-volume *Abrégé de l'histoire naturelle des insectes.*[98] He promised to be less prolix than Réaumur without being frivolous or superficial. "I will take care to say nothing but things that are certain, well attested, and of which I have myself verified a very great number."[99] Though he adopted a lighter tone than Réaumur or Trembley, Bazin also aspired to recruiting observers to look at nature for themselves. Eugène, the male character in the dialogue, claims to have spent the winter observing the astonishing polyps. "I observe day and night and I see prodigies. I advise you, Clarice, to leave aside your domestic duties, to lose your lawsuit, to immerse yourself instead in your aquaria, to collect polyps, and to see the most surprising spectacle ever presented to the human eye."[100] He tells her where to look for polyps and especially how to keep them where she can watch them. Once she has gathered water weeds with polyps attached to them, she is to put them in her dining room:

> You will put these plants in a large glass vase full of water, in a round glass bowl [*cloche à melon*], for example, that you will keep on your table. It will be a little pond for them, they will live there as they do in your ditches, and there you will be able easily, and at your convenience, to contemplate them, to study them, and to do them the service, if it amuses you, of cutting them into pieces.[101]

Eugène discusses various methods for feeding the polyps and watching them digest their meals, and gives Clarice ideas for various experiments she might try. She might feed them different kinds of insects, as Trembley did, and cause them to change color; she might feed a budding offspring to see the food travel through the young to the mother, to be digested and sent back. She might also want to look at the membrane with a microscope. Finally, he instructs her on how to cut the polyp, advising her to wear a leather glove and cut the creature on her fingertip. "Mince it fine, if you wish, like meat for paté, it is all the same to the polyp."[102]

The spectacle of the polyp fit easily into the domestic space of dining room, kitchen, and garden; it could be seen by anyone who learned to pay attention and look closely. This made it rather different from its contemporary rivals in the genre of fashionable spectacles: the effects produced by an electrical machine, say, or the musical automaton Jacques Vaucanson displayed in Paris around the same time. Bazin's book, like Trembley's, asked its readers to expand their attention beyond the page to recreate the phenomena described there. If and when readers took that step, they were also learning to live the life of a naturalist, as modeled by the fictional interlocutor Eugène and, behind the screen of anonymity, by Bazin himself.

6

A Spectacle Pleasing to the Mind

NATURAL HISTORY ON DISPLAY

The naturalist who sacrifices methodical arrangement to visual pleasure forgets that it is not the eyes, but the mind, that he should seek to please; he should not try to offer a spectacle that becomes frivolous as a result of being too pleasing.[1]

As WOULD-BE NATURALISTS LEARNED to look closely at nature under the tutelage of authors like Bazin and Réaumur, they brought polyps, aphids, frogs, and caterpillars indoors in the hopes of seeing elusive phenomena like regeneration or metamorphosis. These pursuits often went hand in hand with acquiring and displaying preserved natural specimens. "The taste for making collections of insects is gaining ground every day," Réaumur told his readers. "People like to see assembled into a cabinet all the insects that curious and attentive eyes can find in the countryside only by searching for them in different seasons, and even in different years."[2] This comment nicely captures the way that displays in a cabinet refracted the contingent and variable experience of observers intent on "seeing and seeing again." Arrays of preserved specimens represented nature's abundance and variety, to be sure, but for those who did their own collecting they also preserved the remnants of excursions to the countryside or the garden, and tangible evidence of their skill and effort as observers. The more extensive the collection, the more people were involved in building and maintaining it, and the more complex the process of pursuing the ideal of shrinking all of nature into the space of a few rooms.

Collections of naturalia were everywhere in eighteenth-century Europe. Sometimes combined with fine art collections, sometimes with antiquities, natural history cabinets could be admired and studied in cities and provincial towns, in private homes, abbeys, medical schools, and botanical gardens. Thousands upon thousands of stones and fossils, stuffed birds and mounted butterflies, preserved reptiles, animal skeletons and shells, human body parts, and wax anatomical models filled cases and cabinets of all sizes and descriptions. Young men traveling around Europe on their Grand Tour—many of them future collectors—visited famous cabinets wherever they went. In Paris alone, there were dozens of collections worth seeking out.[3] Proprietors included aristocrats of the highest rank as well as merchants, apothecaries, and academicians; the collections themselves ranged along a spectrum from "curious" to learned. One chronicler of the Paris collecting scene, Antoine-Joseph Dézallier d'Argenville, distinguished between strategies of display adopted by connoisseurs (*curieux*) and by naturalists, the former aiming to please the eye, the latter organizing their specimens by natural families and classes. D'Argenville was himself an avid collector of shells and knew very well that display cases usually reflected a mix of aesthetic and systematic considerations; indeed, he recommended that serious collectors engage viewers by combining the two criteria.[4]

In the 1740s, afficionados looking for the most extensive Parisian natural history collections, with the best scientific credentials, would head first either to the Jardin du roi, where the royal cabinet was curated by Louis-Jean-Marie Daubenton under Buffon's direction, or to Réaumur's home, where he displayed his own collection.[5] Together, these two sites came to define the field of collecting in the French capital at midcentury, and anyone serious about natural history would visit both places, even though Buffon and Réaumur were known to be personal enemies and at times competed for correspondents and specimens. In practice, the two operations were very different, one a royal institution historically dedicated to public lectures and training medical men, the other a personal collection in a private home. In each case, personal ambition, credit, patronage, and scientific achievement were tied up with rival views of natural history. As Buffon lobbied for royal funds to finance the expansion of a collection that he used as the justification for his major literary work, Réaumur worked on amassing as many birds as possible, a category virtually absent from the royal collections. Although some correspondents sent items to both places and visitors routinely went to see both collections, it seems that Réaumur quite consciously positioned his operation as a contrast to what he saw as its rival, in content and to some extent in style. This chapter looks at how viewers and visiting naturalists experienced Réaumur's collection, what it meant to Réaumur and his collaborators, and what went into keeping it, as it were, alive. Like the "menagerie" of living insects and the gardens where beehives and poultry were kept under observation, the collection functioned as an organic part of the scientific household, linking the household to the rest of the city and to the wider world.[6]

RÉAUMUR'S MUSEUM

In its day, Réaumur's collection was famous for its range and depth, and for the innovative methods of specimen preparation on view there. Visitors came from near and far to see the material record of his natural history practice and to learn how to build their own collections. Although these visitors included aristocratic connoisseurs as well as fellow naturalists, Réaumur took pains to distance himself from the lively market in naturalia that supplied many Parisian cabinets. Indeed, none of the specimens on his shelves had been purchased: "He willingly says this to everyone who comes to visit, and if he does not say it, the labels attached to all pieces will make it clear enough who supplied them."[7] The labels proclaimed the provenance of specimens—not only their geographical origin but also the identity of the sender, often someone who had been a guest in the house. While the arrays of specimens put nature on display, their labels mapped out the personal relations connecting naturalists, travelers, colonial officials, provincial physicians, and anyone else who may have contributed a single item or a whole series.[8] To a visitor reading the labels, geographical origin and the identity of donors became essential to the experience, and the value, of the collection. Unlike most aristocratic collections, the value was not monetary but grounded in the intangible credit of the named donors and of Réaumur himself.[9]

The specimens also made visible various kinds of scientific work, since many of them were the objects of the investigations that filled Réaumur's books and academic papers; when he presented his results to the academy, many in the audience would have been familiar with his cabinets. In many cases, specimens also put on view the preservation methods developed in the laboratory installed in a nearby room. In the early days, as the insect collection multiplied, experiments with different sorts of varnishes, preservatives, containers, and seals occupied Réaumur and his assistants.[10] Later, when the focus shifted to birds, new modes of taxidermy were on display. The collection thus showcased their inventions and techniques, as well as the subjects of their research and the fruits of Réaumur's correspondence.

Réaumur started collecting minerals, fossils, shells, and insects before he had much space for storing and displaying them. When he moved to the Hôtel d'Uzès in 1728, he devoted several rooms to a growing collection that soon included quadrupeds, fish, and reptiles. The collection moved twice thereafter—

first to rue neuve Saint Paul, and then out beyond the city walls to a gracious house at the far end of rue de la Roquette, in the Faubourg Saint-Antoine, where the study of birds soon engulfed both the collection and the laboratory. From the early 1730s, Réaumur's collection was well known to naturalists, travelers, and other collectors. Edme Gersaint, an enterprising dealer in fine art and naturalia, listed it in a guide to local collections, describing it as one of the major attractions on the Paris scene, an essential stop for savants and connoisseurs alike. "Although [M. de Réaumur] has treated this matter as an accomplished natural philosopher," Gersaint remarked, "he has assembled a series [of shells] that will attract the attention even of simple *curieux*."[11] (These afficionados were Gersaint's customers and the intended audience for the catalog in which this guide appeared.)

Although he admired the shells, what made the collection unique, in Gersaint's estimation, was a "complete series" of French metals, minerals, and soils, with their counterparts from other places in Europe, gathered "with extreme care, according to the orders of the late Regent."[12] With this reference to the regent, Gersaint signaled the prestigious pedigree of the mineral collection. Early in his career, Réaumur had supervised what was known as the Regent's Inquiry, a survey of the kingdom's natural resources and industries undertaken by the Academy of Sciences at the behest of the Duc d'Orléans, then regent of the kingdom.[13] Answering the call for information of all kinds, provincial officials sent a steady stream of reports, drawings, and samples of minerals, ores, crystals, plants and plant products, artisanal productions, and occasionally even insects—anything relevant for "the knowledge of nature and the perfection of industry [*les arts*]."[14] As coordinator of the whole operation, Réaumur reviewed the letters and boxes as they came in, analyzed ores and soils in the academy's laboratory, visited mines and other sites around the realm, and corresponded with the intendants and other informants.

The academy had no space for the accumulating boxes of specimens and samples, and a good many of them—especially mineral ores, stones, fossils, soil samples, and shells—ended up on Réaumur's shelves.[15] These became material for research presented to the academy; he boasted in a paper on the fossil-rich soil of Touraine that "there is hardly any province of the kingdom that has not supplied my cabinet with some [fossils]."[16] Years later, when he was looking for a French clay to approximate the material used in Chinese porcelain manufacture, he had only to search his mineral cabinets, where he found kaolin samples for the necessary chemical tests.[17] So the ranks of rocks and fossils materially represented several things at once, displaying the mineral wealth of the kingdom and reminding viewers of the royal interest in improving French industry and prosperity. Réaumur was the custodian of quantities of material generated by requests sent through official channels, backed by the authority of the regent and the academy; almost by default, this became his personal collection, while still retaining the aura of its royal origins. When he showed off these items in his home, Réaumur highlighted his proximity to the crown's prestige and power, as well as his chemical and mineralogical expertise. And as it became more widely known, the collection attracted gifts, some of them spectacular, like a lump of gold from the Potosi mine in South America given to him by the wealthy postmaster-general Pajot d'Onsenbray.[18] Travelers sent stones from the Indies, correspondents obtained samples from mines in the Harz mountains and in Sweden, and former guests sent fossils from Italy and many other places.[19]

The other striking feature of the collection was the vast number of insects, some dried between sheets of glass and some preserved in spirit of wine. Both Gersaint and d'Argenville mention explicitly, as a remarkable novelty, that "all the different states" of each insect were laid out together. Butterflies and beetles, long a staple of curiosity cabinets for their color and variety, were reclaimed for science by appearing wherever possible alongside their own caterpillars, nymphs, and chrysalises; many of these had been cultivated in the living collection or gathered in the gardens just outside. Insects not suitable for drying, like worms, sowbugs, roaches, wasps, flies, and mosquitoes, floated in different sizes and shapes of glass tubes and vials designed for the purpose, the "very novel inventions" of the collector.[20] Inventions proliferated as the collection expanded into other categories of animal, requiring more attention to maintenance of the pickled specimens and new ways of mounting dried birds. The number of correspondents contributing to the cabinets grew over time, so managing the collection also meant managing correspondence—soliciting specimens, instructing neophytes about tricks for successful packing for transportation, sending supplies and instruments, tracking down lost shipments, thanking correspondents when specimens arrived in good order, and repaying them with various favors and services in the capital. The objects in their cases embodied and crystallized, so to speak, a multitude of practices: experimentation, dissection, organization, communication, and preservation.

THE COLLECTION IN RUE DE LA ROQUETTE

In 1740 the cabinets were packed up for the move to more spacious quarters in a sparsely developed neighborhood just outside the old city walls. "The extent and beauty of the gardens of my residence," Réaumur wrote to Pierre Baux

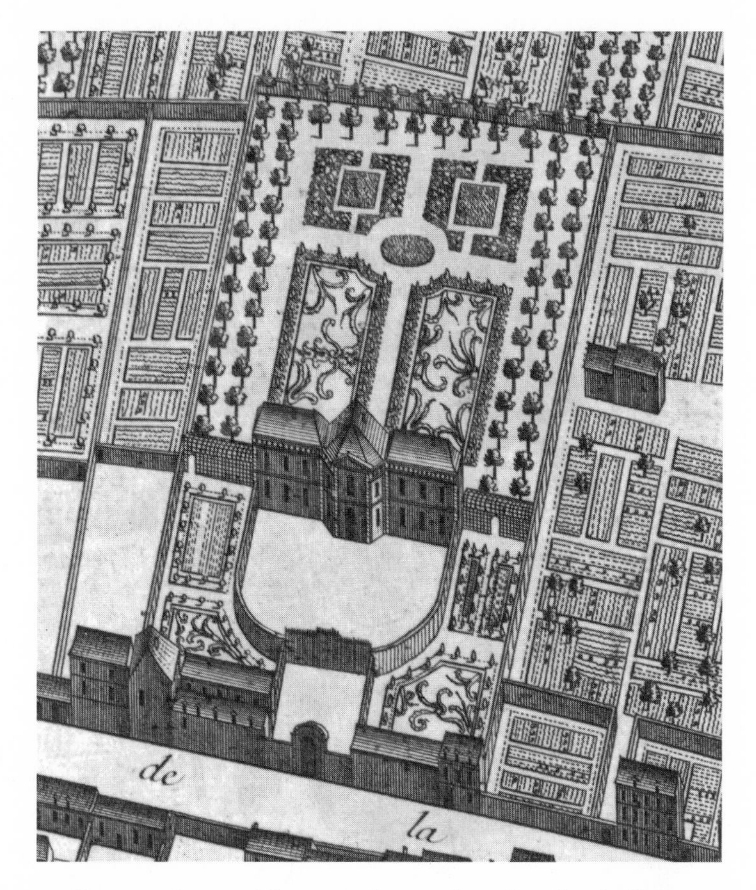

Figure 6.1. Réaumur's house, gardens, yards, and ancillary buildings, rue de la Roquette. Louis Bretez, *Plan de Paris* (known as the Turgot map), ca. 1739, plate 5 (detail). Harvard Map Library.

✲

in Nîmes, "the convenience of the accommodations for my cabinets and my laboratories, determined me to flee the center of Paris, where after all it is the task of my horses to take me every day."[21] The Faubourg Saint-Antoine was a neighborhood where the workshops of artisans—glassblowers and metalworkers and cabinetmakers—gave way to fields and comfortable houses built for wealthy owners as retreats from urban life.[22] After the move, Réaumur gave up the lease on his riverside house in Charenton, consolidating laboratories and collection in one location, with the advantages of "both town and country."[23]

The house and grounds on rue de la Roquette can be identified on the "portrait-map" of Paris by Louis Bretez, and Jacques-François Blondel included a detailed description of the features of the house in his book on French architecture.[24] Opening off a formal forecourt were enclosed yards, or *basse-cours*, "equipped with the buildings that supply the necessary appurtenances for an edifice that can be regarded as a country house [*maison de plaisance*], being situated at one of the extremities of Paris." These ancillary buildings included the "orangerie house" and other houses along the street, where most of the residents of the household must have lived. The main house divided the front courtyards from the extensive gardens at the back; its second floor was well supplied with windows to illuminate the collections, which gradually expanded to occupy most of the upstairs rooms, originally designed as living quarters. By 1752, when Blondel visited, Réaumur's bedroom was the only room upstairs that was not given over to natural history specimens. On the ground floor, connected to the bedroom by a staircase, he had a large study or workroom. The antechamber off the central vestibule was turned into the library, "this house in general being too cramped for the savant who inhabits it."[25] Guests were received in the other large rooms, a salon, a drawing room, and a dining room. Many kinds of people came to admire the collection, whether travelers passing through Paris, friends invited for dinner, or colleagues with a more specialized interest in some part of the collection. Although it did not have regular hours (as did the Jardin du roi), Blondel remarked that Réaumur's museum was "made available to the public with an obliging willingness and an affability worthy of the zealous citizen who owns the collections, . . . the most complete in Europe."[26] Anyone who asked would be shown around, either by the proprietor or by one of his assistants or associates. We have already encountered Jean-Antoine Nollet showing the Provençal naturalist and bibliophile Jean-François Séguier around the insect and mineral collection in the early years. Other references to such visits are scattered throughout the correspondance. In 1752, visiting Paris as tutor and chaperone to the young Duke of Richmond, Trembley introduced his charge to intellectual and social life in the capital. Trembley was by this time an old friend of Réaumur, and the visitors spent several convivial evenings in the Faubourg Saint-Antoine, viewing the collections and dining with a group of academicians, several of whom had worked in the collection. "In a first visit, the Duke of Richmond saw the birds at M. de Réaumur's house, shown around by the host himself," Trembley reported to Bonnet. "Yesterday M. Bernard de Jussieu showed him the fossils and started on the minerals."[27] Father Lignac, another intimate friend of the Réaumur household, was also an enthusiastic spectator on his visits to the capital:

> Several times I had the pleasure of admiring [the collections] on those days when there was a press [*affluence*] of foreigners. Everyone was as moved as I was

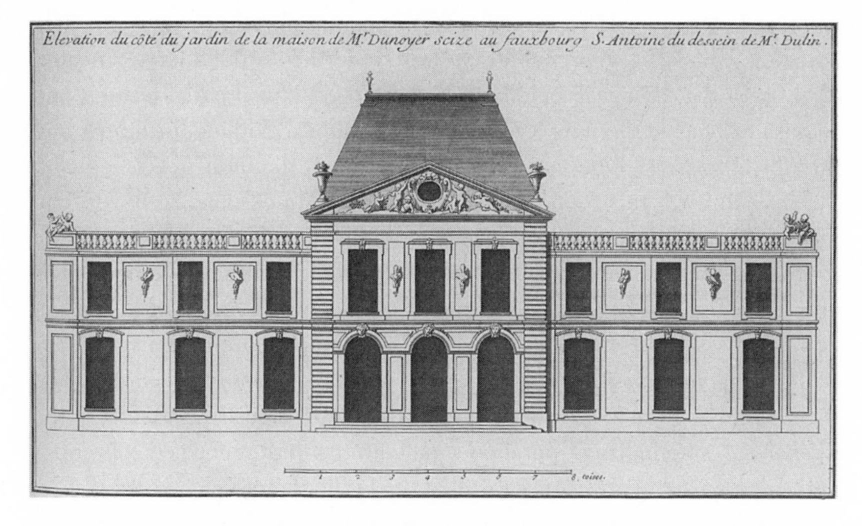

Figure 6.2. Réaumur's house in Faubourg Saint-Antoine, as viewed from the garden. Jacques-François Blondel, *Architecture françoise*, vol. 2. Typ 715.52.219, Houghton Library, Harvard University.

✻

> myself by the general impression made by the immense quantity of different objects, and everyone was equally struck by the order in which they were placed.[28]

Specimens were placed according to their "relations" to each other, determined by either functional or anatomical similarities. Lignac implied that the order was self-evident to viewers, who were meant to be instructed rather than dazzled with symmetrical designs and pleasant vistas.

The most detailed description of the layout and contents of the whole collection comes from the inventory made as it was about to be dismantled, after Réaumur's death.[29] The inventory is a dry official document, with only cursory descriptions of the contents of the cases, bottles, boxes, armoires, and shelves. Nevertheless, it allows us to follow the notaries as they moved systematically through the rooms, and gives a sense of the scale and layout of the collection as a visitor would have encountered it in the 1750s. Before going further into the acquisition of the specimens and the personal relationships that made possible their accumulation, I draw on the inventory and corroborating documents to recapture something of the experience of someone walking through the collection for the first time. Climbing the staircase to the first floor, the visitor entered a large room filled with all sorts of dried and stuffed animals, many of them from distant corners of the world. Five seal skins lay on the floor, and all sorts of animal parts vied for space on the walls: the horns of ibex, rhinoceros, and South African wild cows, jawbones of boar and horse, a camel stomach, a human stomach, two armadillos, a lizard, and the skin of a large anteater. Shelves on the same wall held bottles and jars of animal parts, fetuses, and "monsters"; facing the door a selection of "weapons, clothing, furniture, and utensils of savages" hung next to a zebra skin and the fetus of an elephant. Strolling through this room, a curious guest could peer into no fewer than fifty-eight glass-covered cases containing dried and stuffed quadrupeds, from roedeer, hedgehogs, rats, ferrets, rabbits, and foxes to guinea pigs, various kinds of monkeys, stoats, beavers, and lemurs. Introducing the range of the collection's contents, this first room established it as much more than a specialist's cabinet, rich in curiosities as well as more mundane series of specimens. The sheer variety—from local insects to crocodile skins, monstrous fetuses to the weapons of Amerindians—put it on a par with the most renowned collections of the day, including the royal cabinet at the Jardin du roi.[30]

If the first room contained many objects and animals that might have been seen in any respectable natural history collection, the next three presented a striking array of hundreds of stuffed birds the viewer would be unlikely to encounter anywhere else. These were the signature preparations that Réaumur and his assistants had spent more than a decade perfecting, with the help of numerous correspondents. Entering the second room, "No one could be insensible to the impression [*coup d'oeil*] that the series of birds of this rich collection provides."[31] Individual glass display boxes protected the specimens from vermin and allowed them to be shifted around to accommodate new arrivals. Some cases contained familiar birds like magpies, crows, jays, plovers, bustards, and spoonbills; others held cormorants, herons, gulls, and storks; then there were exotic toucans, birds of paradise, cranes, cassowaries, parrots, and parakeets. Still other cases displayed varieties of poultry and game birds: pheasants, guinea fowl, pigeons, doves, chickens, quail, and partridge. Perhaps the most impressive specimens in the bird room were a peacock and an ostrich, each in its own large case.[32] The large bird room gave onto two smaller rooms. One of these held thirty-five cases of birds of prey and fifty-nine of river birds; the other, in addition to another hundred glass cases for individual birds, had several large armoires whose shelves were packed with nests and eggs.[33]

Continuing through the rooms of birds, our traveler could venture into a little alcove filled with dozens of shelves packed with jars of seeds, fruits, bark, and leaves, and two small armoires displayed gums, resins, and sap—all from foreign plants. Retracing his steps, he would pass by the quadrupeds again and on into the other side of the house, where a large room housed the vast min-

Figure 6.3. Hélène Dumoustier, bird nest from Réaumur's collections. Archives, Académie des sciences, Fonds Réaumur, dossier 10. By permission of Archives de l'Académie des sciences, Institut de France.

✲

eral collection, including fossils, samples of building stones, stalactites, coal, amber, ores, metals, and so on. This part of the collection included the samples gathered years before for the Regent's Inquiry, as well as other items, especially fossils from all over Europe.[34] The inventory mentions soils and sands of all types, as well as ores and minerals from China, Saxony, Siberia, and Sweden. No fewer than seventeen armoires filled with specimens lined the room; one wall was taken up with two large chests holding a total of ninety-six drawers of various sizes. Above the armoires, mounted on the walls and above the doors, were models of machines, such as mechanisms for lifting boats through canal locks, and samples of exotic clothing, including "a complete outfit of Chinese clothes, enclosed in a box."[35]

In the middle of the large room, the fifth described in the inventory, stood a bureau of black wood containing forty-six drawers filled with dried insects. Opening the drawers, the curious visitor found hundreds of little boxes with glass tops and bottoms, giving clear dorsal and ventral views of "insects of all kinds, like butterflies, including some foreign ones, and fireflies, and others with the silk cocoons belonging to those insects."[36] This was just a preview of the insect collection, which continued into the drawers and shelves of the next room. Here were the insects familiar to the readers of Réaumur's books: grasshoppers, dragonflies, fleas, gall insects, all kinds of beetles, caterpillars, spiders, centipedes, scorpions, mites, "and others." Some of the glass boxes held dissected specimens, mounted to show anatomical features. One cabinet contained glass tubes with insects and reptiles preserved in spirit of wine; in addition to caterpillars, there were snakes, chameleons, salamanders, and lizards. Mounted on the walls were dried specimens of larger reptiles like crocodiles and iguanas, as well as ants' nests and wasp nests from Cayenne and other places mounted alongside samples of the wood used by wasps to make their nests.

Finally, entering from the insect room, our visitor came to a small cabinet devoted to sea creatures. Dried fish hung from the ceiling, including a hammerhead shark and a large swordfish. Above the door, along the windows, and in the doorways he could spot such curiosities as the bones and teeth of whales, three narwhal horns, a dried sturgeon, and tortoise shells. The cases and shelves and drawers in this room housed the substantial shell collection noticed by Gersaint, and myriad glass boxes containing sea stars, corals, madrepores, sponges, sea urchins, mollusks, and other marine animals. As with the insects, some were dried, some floated in spirit of wine. A stroll through the sequence of cabinets was a tour through the three kingdoms of nature; it was also a tour of many of the subjects investigated by Réaumur and his associates over the years. Walking through the rooms, spectators came face to face with the material record of the collector's dense and productive network of correspondents, and the tangible results of many years of experimentation and observation. Thus, in addition to presenting a tableau of the prolific variety of nature, the cabinets put on view a record of work and pleasure, inventions and techniques, sociability and exchange.

The notarized inventory, with its tabulation of the different kinds of containers and pieces of furniture, and rough indication of their contents, registered a static snapshot of the famous museum as it looked at the time of the proprietor's death, with no indication about the origins or meanings of the specimens. What went into amassing and maintaining such a collection? This

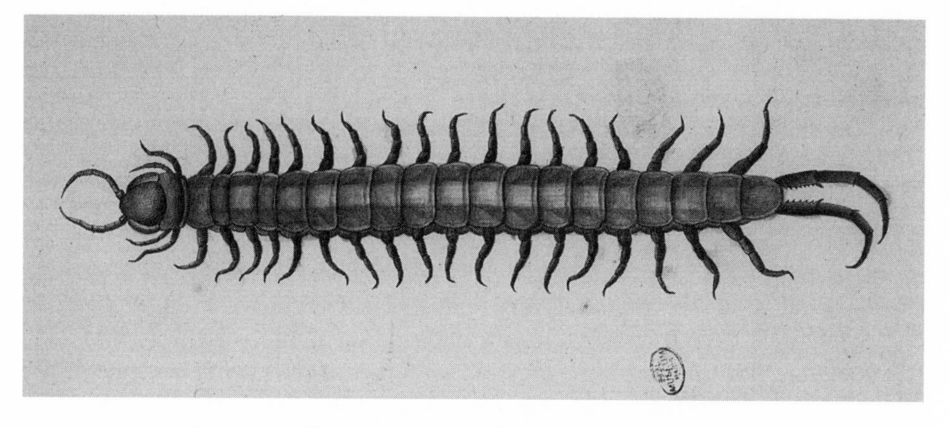

Figure 6.4. Hélène Dumoustier, millipede from Réaumur's insect collection. Ms. 972, BCMHN. By permission of Bibliothèque centrale du Muséum nationale d'histoire naturelle, Paris.

❋

question leads us into the life of the collection itself, as it grew and developed through the efforts of the residents of Réaumur's household, his friends and associates in Paris, and his far-flung correspondents and various go-betweens. Another inventory, made by Guettard when he as living in the household in the 1740s, provides further information about the provenance of many items. Probably working from the labels, Guettard included names of contributors, which can be corroborated from correspondence in many cases. Though it is not complete, this document gives more details than the notarial inventory about some of the objects.

The bulk of the collection, as we have seen, came from French fields and forests. Alongside these more or less familiar creatures were foreign specimens sent from all parts of the globe. Some were spectacular, like the ostrich skin, beetles from Madagascar, flamingos from Avignon, and a lizard with a double tail sent from Rome by Father Mazzoleni.[37] There was even a bear sent from Canada.[38] Some of the rarest items carried the impeccable credentials of their donors: a faun with two heads sent from Fontainebleau by the king; a cardinal from Mississippi from the Princess of Conti; Brazilian birds from the Prince of Craon; a pelican sent via Nollet by the Duke of Savoy.[39] Other objects we have encountered in previous chapters: delicate anatomical preparations of caterpillars mounted by Bazin; pine caterpillar nests sent from Bordeaux by Raoul; marine creatures collected by Richard and sent from his tanks near La Rochelle; corals and sea polyps gathered by Guettard in Normandy. Though no longer alive, of course, these things were given a kind of afterlife as part of the continually shifting arrays of specimens, dynamic elements in an expansive matrix of research and sociability.

The tentacles of the enterprise in rue de la Roquette stretched out to learned and public audiences at the Academy of Sciences across the city; they drew visitors to the Faubourg Saint-Antoine to tour the cabinets; and they generated further activity around France and around the world via letters and boxes and books. Traces of Réaumur's friendships and patronage relations survive in personal letters documenting shipments and exchanges of favors and techniques of preservation. During his lifetime, these traces would have been visible to visitors in the objects themselves, in their labels, and in the spoken commentary of the host or one of his resident assistants. The labels commemorated personal friendships, and Réaumur often referred to these inscriptions as markers of recognition and appreciation, as in this letter to Pierre Baux:

> The bird of prey that you took the care to send me has arrived in very good condition; it is now in its place in my cabinet, where an example of its kind was lacking. Next to it is a label giving the name it bears in your region, and . . . its Latin name. This same label says that I got it from you, and perpetuates the acknowledgment of my appreciation, which is proportional to your generous wish to enrich my collection further.[40]

Or, to the Swiss mathematician Jean-Pierre de Crousaz, who also collected minerals: "I already have in my cabinet some pyrite crystals with regular faces . . .; nevertheless I will put yours there, if only to have the pleasure of seeing from time to time a label bearing a name I respect so much."[41]

The bird collection depended as much on experiments with preservation and mounting techniques as on the goodwill of distant correspondents like Baux. A series of assistants, starting with François-David Hérissant, designed and carried out a continual stream of experiments related to preservation. These ranged from designing seals and stoppers to prevent evaporation from jars of pickled specimens, to testing the means of drying birds and quadrupeds, to exploring the effects of different salts on carcasses and the effects of spirits and other distillation products on the colors of feathers. Réaumur and Hérissant presented some of the results to the Academy of Sciences; others were reported directly to distant correspondents.[42]

"No cabinet," Réaumur wrote to a friend in 1744, "gives such a pleasing and amusing view [*coup d'oeil*] as one filled with birds that the countryside, woods, swamps, and rivers can only show us successively."[43] Though the spectacle of hundreds of birds in lifelike poses impressed his visitors, Réaumur imagined

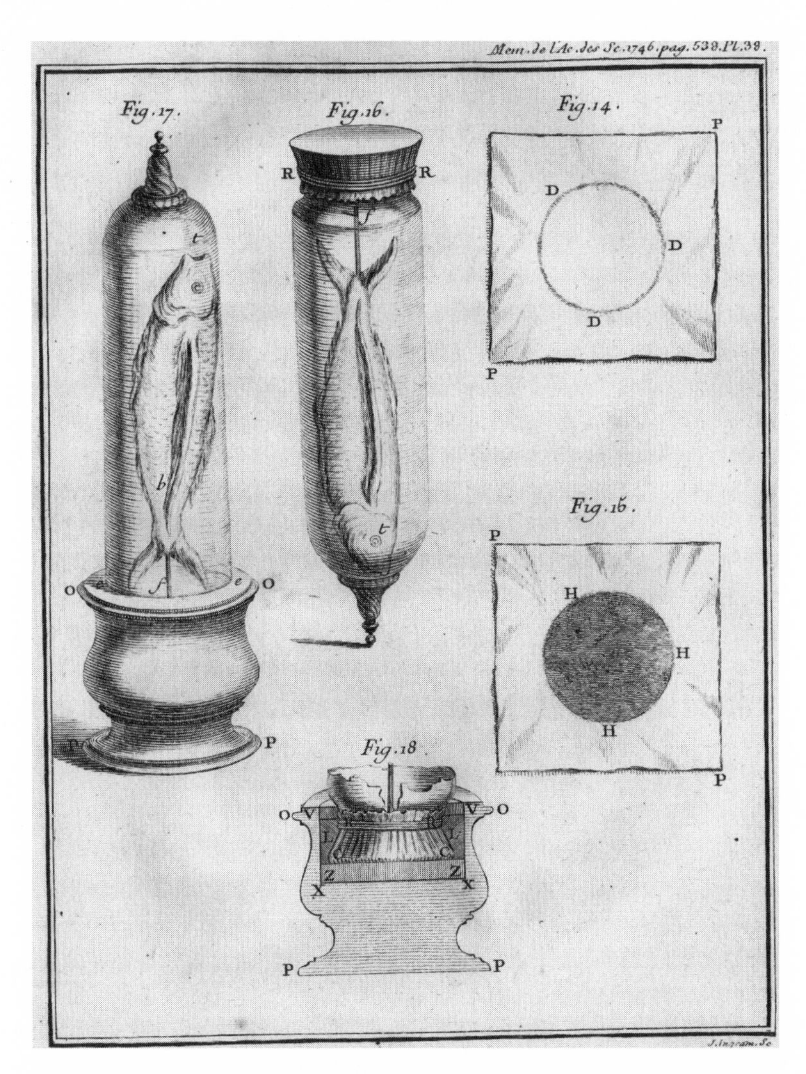

Figure 6.5. Preservation and display in Réaumur's collection: technique developed by his assistants for sealing jars to prevent evaporation of *eau de vie*. Réaumur, "Moyens d'empêcher l'évaporation des liquides spiritueuses, dans lesquelles on veut conserver des productions de la nature de différens genres," *Mémoires de l'Académie royale des sciences* (1746), plate 38.

✡

that they would also contribute to the "progress of ornithology."[44] Better means of preserving and mounting would allow collectors to bring together "more different kinds . . ., for convenient examination and comparison, than one could hope to find successively even in the longest and most difficult voyages."[45] Réaumur's letters in this period are full of comments about the dearth of well-preserved birds in collections and the limitations of existing methods of taxidermy. "For a long time I have regretted that no one knew any other method for preserving dead birds than stuffing them [with straw or other materials], which demanded time and skilled hands and did not give sufficiently perfect or sufficiently durable results."[46] Before he started collecting birds in earnest, Réaumur and his assistants conducted numerous trials of methods for drying and mounting the carcasses. "By degrees," he told Trembley, "I succeeded in preserving an appearance of life in dried birds; to make them look just as they were when they were ready to fly or to walk."[47] Drying the bodies in the gentle heat of an oven, without even eviscerating them, gave the best results, and the technique had the added virtue of simplicity. This "simple art" could apparently be learned in "less than half an hour"; a single person, he claimed, could prepare thirty or forty such specimens in a day's work. He boasted to one of his correspondents that in a single year of practicing the simple methods invented and tested in his laboratory, he had assembled in his cabinets more than six hundred birds, an array that never failed to impress viewers. "They are all in the attitudes natural to them; in a word, at first glance they appear to be alive. They are nothing like those stuffed birds whose different parts no longer have their true proportions."[48]

The key to the lifelike poses was a new invention, "an extremely simple machine," inspired by a farrier's trevise, the contraption used by blacksmiths for restraining horses while they were being shod. After fastening the bird to the device, a kind of wooden frame mounted on a flat base, "one can, in very little time, position the bird with wires and bands [tied to the frame] in whatever of its natural attitudes one wishes."[49] The bird's feet were first stitched to cards, with threads around each claw, and then these cards were pinned or nailed to the wooden base. For bigger birds, strong ribbons supported the body, passing under the wings and tied to the crosspieces of the frame; for smaller birds, lighter bands or strings would suffice. A flexible wire inserted into the body through the eye served to position the neck and head. Once arranged, the operator put the whole contraption, with bird attached, into the oven to dry.[50] Although the process required some dexterity, the bird did not have to be eviscerated or skinned, reducing the risk of damaging the carcass.

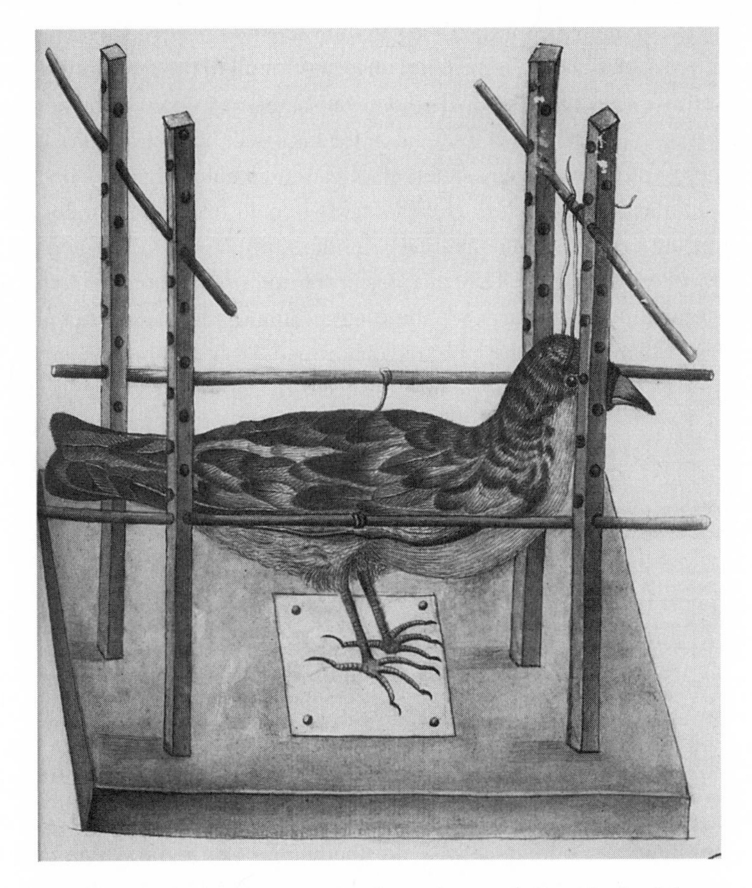

Figure 6.6. Hélène Dumoustier, device for immobilizing birds while drying. Ms. 972, BCMHN. By permission of Bibliothèque centrale du Muséum nationale d'histoire naturelle, Paris.

✲

As the work on specimen preparation flourished, Réaumur planned to add a chapter on preserving birds and quadrupeds for collections to his unfinished seventh volume of insect studies on beetles and ants. By this time, he and his associates had a set of protocols for protecting dried and stuffed specimens from the ravages of mites and other insects. "The hundred little details necessary for complete success would fill quite a long *mémoire*," he wrote to Trembley, who had inquired about how to preserve specimens.[51] Gradually, collection and preservation took precedence over finishing the insect manuscript, and the projected chapter on preservation ballooned into a book-length manuscript. By early 1745, so many people were asking him how to preserve specimens that he planned a small book "to respond once and for all to the questions that are repeated to me so often, and which I cannot answer to my satisfaction even with very long letters."[52] Just as he intended his insect books to inspire readers to observe, cultivate, and collect insects, Réaumur wanted anyone with "an inclination to make natural history cabinets" to benefit from his laboratory trials of taxidermy techniques and materials, "for their private satisfaction and for the progress of ornithology."[53] The fruit of years of experimentation on methods of drying, embalming, stuffing, and mounting specimens, the manuscript covered the preservation and display of birds, quadrupeds, fish, and insects, including a chapter on the different species of insects that could destroy the mounted specimens.[54] In the event, the how-to book on collections never made it to the press, possibly because of delays with producing illustrations. With the larger publication project in limbo, Réaumur printed a pamphlet version of four pages, with compact descriptions of four alternative methods for preserving and packing specimens for shipments over long distances.[55]

Stepping up the acquisition of birds and animals from distant places, Réaumur sent this little brochure to "all parts of Europe and all parts of the world," enlisting the help of correspondents and other contacts and connections throughout the French bureaucracy. He asked Trembley, whose connections extended into the English and Dutch elite, to find a way to get the pamphlet onto Dutch ships traveling to the Moluccas, so he could get birds "that we cannot get from French vessels." Perhaps he could ask his old patron Count Wilhelm Bentinck, a diplomat with substantial interests in the Dutch East Indies Company, to put in a good word.[56] Trembley quickly translated the text into Dutch and had it printed in Holland; with the support of Bentinck it went out on Dutch East Indiamen. Trembley, brokering the relations between his social superiors, worked as a crucial intermediary in the complicated patronage economy connecting him to both Réaumur and Bentinck. He produced the Dutch-language version of the pamphlet without being asked and refused to be reimbursed for expenses, insisting implausibly that the printer had produced the pamphlet without charge.[57]

A slightly revised version of the pamphlet was printed in 1747; once again Réaumur sent a packet to Trembley for distribution to the Indies. This version was translated into English as well, and published in the Royal Society's *Philosophical Transactions*.[58] Michel Adanson, a young naturalist who prepared himself for an expedition to Senegal by immersing himself in Réaumur's collections, carried a copy of the pamphlet with him to Africa, from where he successfully sent many exotic birds to Paris.[59] The text—without benefit of illustrations, which surely would have made the condensed instructions easier to follow—encouraged readers to send birds or other animals using relatively simple techniques to keep carcasses from rotting or falling apart en route. These techniques had all been tested in Paris; the reader was assured, for instance, that though birds immersed in brandy seem to lose their color over time, "repeated experiments have nevertheless taught M. de Réaumur that the color of feathers is not affected by the strongest brandy and even spirit of wine, and that after one has let the bird . . . dry, one can easily put the feathers back into their natural state, and make it look as it did in life."[60] It took considerable effort to carry out the numerous "little attentions" and precautions necessary to ship specimens successfully. Indeed, as we shall see, many specimens packed with the best of intentions did not arrive in Paris in good condition.

The three recommended methods were all suggested as alternatives to the standard practice, difficult for the inexperienced hand, of stuffing skins with straw or other soft material. The techniques perfected in Réaumur's laboratory included soaking carcasses in spirit of wine or strong brandy, drying in an oven with or without the frame support discussed above, and embalming or salting with various chemicals or spices. Each of these was conceptually simple but in practice required an ability to adapt to contingent circumstances and to make judgments and choices. The pamphlet gave the traveler (or his servants) several options, to allow for local conditions and available commodities, though in the limited space of four pages it could hardly spell out unambiguous prescriptions for the preservation of all possible specimens. "There are several methods, simple in themselves," Réaumur noted, "for putting and retaining the bird in a natural position, that nevertheless would be very long to explain in detail: the little we will say about it will suffice for industrious people wishing to make use of it."[61] In effect, each shipment turned into a kind of experiment, whose outcome would only be known once the cases and barrels of specimens reached their destinations.

In some cases, travelers found the advice sent from Paris to be altogether useless. Pierre Poivre, who traveled widely in Asia and contributed some of the most exotic species in Réaumur's bird collection, pointed out that conditions in remote locations made improvisation necessary:

> You will be surprised that I kept to my old method for the conservation of birds, and that I did not follow the one you told me about in your last letters. But you will excuse me when you learn that this method is impracticable especially in the

> countries where I made my collections, where one can find neither barrel nor cask, neither alum nor salt. It is difficult to have the least article for the necessities of life, and our colonies lack absolutely everything. . . . So I did the best I could and you will see that the most beautiful birds kept their colors very well.[62]

The many people who took up the challenge to observe, collect, and ship specimens do not fit a single or consistent profile. One eager contributor was Jean-François Séguier, a botanist, bibliophile, and antiquarian.[63] Originally from Nîmes, Séguier had traveled all over Europe as companion and secretary to the Italian nobleman Scipion Maffei, a renowned antiquarian and collector of ancient medals and inscriptions, before settling for twenty years with his patron in Verona. On their Grand Tour in the 1730s, they had spent several years in Paris, where they became well acquainted with Réaumur, his social circle, and his collection in the Hôtel d'Uzès. Subsequently, Séguier maintained his ties to French collectors by sending gifts of fossils from the rich deposits near Verona, in exchange for books from Paris shops.[64] He also volunteered to gather and ship Italian birds not yet represented in Réaumur's cases.[65] An inveterate collector himself, as well as a dedicated scholar, Séguier threw himself into the task with a will, negotiating with local bird hunters, poultry merchants, friends living in the mountains, and neighbors in town. In one case, he knew of a pet song thrush owned by someone in Verona who did not wish to give it up; he proposed "an exchange for something she might like from Paris, which you could easily get for her." Some species were hard to come by, including an Italian variety of grouse (*cedrone*): "It is almost as large as a *coq d'Inde.* They are ordinarily all black, and this is a bird of beautiful demeanor and great pride. Its flesh is very delicate, but it is very rare here."[66] (It took more than a year to obtain one of these, but eventually it ended up in Paris.) As Séguier acquired the birds, he "read and reread" the little pamphlet, mulling over the different preservation techniques, wrapping the birds carefully in linen one by one and immersing them in glass jars of brandy cut with vinegar, to save on the cost. He decided against drying the birds and filling them with alum, and instead packed the individually wrapped specimens into a barrel, cushioning them with layers of flax and filling the container with fresh vinegar before sending it off to Paris, via his booksellers in Geneva. The tasks associated with accumulating, storing, packing, and shipping had proceeded by fits and starts over the better part of a year.

When the barrel finally arrived in Paris, after several months in transit, Réaumur was distressed to see that many of the birds were not in good shape, in spite of the attentions lavished upon them in Verona, and he had to break the bad news as delicately as possible. The problem was not with the meticulous packing job, he said, but with the liquid used as a preservative:

> It is not at all your fault, but my own. I told you once about vinegar as a liquid appropriate for preserving birds, without sufficiently explaining myself about the precautions necessary for using it. Subsequently I neglected to point out that, in the printed instructions, I indicate only various kinds of distilled liquors as proper for this purpose. I admit that I was anxious as soon as you told me that . . . you had filled the barrel with the best vinegar you could find. My experiments had proved to me that little birds should not be kept more than several weeks in this liquid, after which they must be taken out and dried.[67]

One can only imagine Séguier's dismay at the description of the sorry state of his birds, with their beaks rendered "as flexible as wet paper" from soaking in vinegar, and the bones of the heads and skeletons so soft that they could not support the structure of the body. Though some of the larger birds were salvaged from "this disaster," the parchment labels attached to each specimen were entirely illegible, the ink having dissolved in the vinegar. Réaumur reiterated his earlier advice to send the carcasses dry, eviscerated and filled with alum, and then wrapped in cloth to keep the feathers in order and protect them from the motions of the carriage.

Mortified, Séguier undertook the substantial task of replacing the ruined specimens, while continuing to gather other species. In spite of the failure of his first shipment, his dedication made him an invaluable source of new birds:

> I had almost foreseen that the vinegar was softening the beaks and bones too much: several times I had spread open one or two little birds to see what state they were in. The beaks, softened a little too much, caused me some distress, but I imagined that you had some methods unknown to me for hardening them, and I mistakenly reassured myself on that score.[68]

He promised to tell his suppliers and friends to hunt for the same list of birds again, and planned to use brandy instead of vinegar as the short-term preservative, before shipping them dry in well-sealed containers. The larger birds, per Réaumur's strong recommendation, he would fill with powdered alum before packing. He planned to try keeping many of the smaller birds alive until he was ready to ship them, to avoid immersing them for too long.

Séguier's surviving letters to Réaumur show the extent of his ornithological operation in Verona, where he willingly added the acquisition and preservation of birds to his many other scholarly and natural history pursuits. To acquire the desired specimens, he had to deal with trappers and hunters and merchants, as well as friends; to process them, he bought various supplies, including chemicals, containers of glass, wood and glazed pottery, and wrapping and packing materials. His work space housed cages of live birds, jars of carcasses soaking in alcohol, and a staging area for eviscerating, drying, and stuffing specimens; his extensive library included the reference works used to identify them. He kept careful records, as attested by the lists of birds and the labeling system he developed, with the key sent to Paris separately from the specimens themselves. Though he had not been a bird collector before, Séguier turned out to be the ideal contributor to the ornithological collection, in that he was already engaged in other branches of natural history.

In his quest for specimens, Réaumur cultivated and exploited, with extreme politeness, all sorts of connections. Some of these, like Séguier, supplied him directly, in a complex economy of exchange, recognition, and reimbursement; others, including booksellers, priests, trading company officials, and aristocratic diplomats, facilitiated the movement of letters and cases outside the normal postal system; and still others used their own contacts to solicit specimens from distant locations. Trembley, living in England, translated and distributed the instruction pamphlet for foreign audiences. He was repaying the service performed for him only a few years earlier by Réaumur, who had lent his authority to the discovery of the polyp's strange abilities. Trembley proved a crucial link to sources beyond the French colonies, notably through Count Bentinck, in whose household he had made the initial observations of the polyp. Both had ties in England, as well as Holland, and traveled back and forth between these places, sometimes via Paris. Bentinck's request to the captains of Dutch East Indiamen and to local officials at the Cape of Good Hope and other places around the Indian Ocean, quickly generated shipments from the all over the Indian Ocean and southern Africa. As Réaumur exclaimed to Trembley when the first such cask arrived in France, "What, some birds from the Cape of Good Hope have already arrived! It seems that Count Bentinck only had to say the word."[69] Unpacking them, he found them in excellent condition, and none were duplicates. "They did not suffer at all from their travels either by sea or by land; one would think them full of life."[70] Whoever had packed the barrel had taken to heart the instructions to wrap each specimen separately before immersing it in liquid. Only a few weeks later, another three large barrels of birds and fish arrived—more than a hundred specimens altogether, from Ambon and the Cape. It took two weeks to unpack, dry, and mount them:

> It is only after one has worked to mount the birds advantageously that one can really see what is missing and if they have lost too many feathers en route. It was

Figure 6.7. F.-N. Martinet, turtledoves from around the world: North America, Cape of Good Hope, and Ambon. M. J. Brisson, *Ornithologie*, vol. 1, plate 9. Wellcome Library, London.

✲

> not until yesterday evening that we finished preparing them and that I had the pleasure of considering them one by one, to see the ones that are not in my collection and that I had so wished to put into it.[71]

Although Bentinck was not shipping the birds himself, acting as a go-between to arrange shipments at a distance, Réaumur reported to him on the condition of the birds and gave advice about improving the results, no doubt hoping that this would be passed on. Given the excellent quality of the specimens arriving in Paris from distant ports, it seems that anonymous collectors, who remained unknown to Réaumur, as to us, took the task of acquiring, preserving, and packing very seriously indeed.

Some of the travelers who visited the collection on rue de la Roquette were inspired to contribute to it after they returned home. One illustrious visitor was James Douglas, Earl of Morton, a Scottish lord on a diplomatic mission to Paris in 1746. Two years later he was sending barrels of sea birds from the coast and islands near his estate in Scotland, something he continued to do sporadically for the next five years. Thanking him for the first selection, which included a pair of Soland geese he had been wishing for, Réaumur noted that his collection "has grown considerably since you saw it."[72] Réaumur's letters (Morton's do not survive) retain echoes of the kind of conversation the two must have enjoyed in Paris, sometimes no doubt in the very rooms where the stuffed Scottish birds would end up. In addition to the perennial topic of shipping routes and packing techniques, they discussed whatever they had heard about particular birds, the necessity for caution and "enlightened eyes" when assessing third-party reports about bird behavior, and the diet of birds with distinctively shaped beaks.[73]

In this period, political circumstances interfered with shipping in the Channel, as well as the Atlantic, and several times no reliable carrier could be found for the barrels Morton had prepared:

> It would certainly be very reasonable, as you suggest, to establish a cartel for everything concerning the sciences. If there were such a thing, I would not have so many complaints about the English corsairs who have taken from me a great number of shipments; fear of them has prevented the shipment of a great many others.[74]

Three years later, it still took months to receive a barrel from Scotland; the birds, immersed in good aristocratic brandy, arrived in perfect condition, even though they had spent as long in transit "as if they had come from the East Indies."[75]

At the height of his collecting activities in the mid-1750s, Réaumur received letters and boxes of specimens almost daily.[76] Some contributors were noblemen like Bentinck and Morton; others, like Séguier, were fellow naturalists and collectors. Yet others were Frenchmen in service to the crown, such as Jacques François Artur, who spent thirty-five years in the Atlantic colony of Cayenne. Artur, a young physician with an interest in botany and natural history, had come to Paris to study at the Jardin du roi in the 1730s, and made the acquaintance of Réaumur.[77] He corresponded with Bernard de Jussieu on botanical matters, sending Caribbean plants and seeds for the Jardin du roi whenever possible.[78] Through the 1740s, he sent insects, birds, and other animals to Réaumur as well. Their letters record efforts and expectations continually frustrated by the tropical environment, the demands of Artur's patients in Cayenne, and the vagaries of transatlantic shipping. To transport specimens of fauna from the tropics into the safe and ordered space of the Parisian cabinet

Figure 6.8. F.-N. Martinet, Scottish hazel grouse (sent by Lord Morton) and quail from Louisiana. M. J. Brisson, *Ornithologie*, vol. 1, plate 22. Wellcome Library, London.

✲

entailed a continual struggle, on both sides of the Atlantic. For every equatorial specimen eventually displayed, dozens of others did not survive the processes of storage, packing, preservation, and shipping.

Réaumur specified that Artur was to send things "to my address and not to that of the Jardin du Roy"; when he received the fruit of these efforts, he promised "to inform Count Maurepas, the academy, and the public, what I owe to you."[79] Maurepas was the royal minister charged with oversight of the Academy of Sciences and the botanical garden, as well as the navy, so Artur tried to use his scientific patrons to improve his position in Cayenne, where he depended on ministerial goodwill for his stipend. He repeatedly asked Réaumur to intercede with Maurepas to obtain more generous remuneration for his service as the only doctor in Cayenne.

Initially, sending his books on insects as gifts for the colonist's library, Réaumur was curious to acquire insects—anything at all, rare or common.

Everything except butterflies could be thrown together into a large bottle filled with brandy and "as much sugar as it can dissolve," the standard recipe for the killing-jar he called an insect "graveyard."[80] Not surprisingly, Artur found it difficult and expensive to obtain French *eau de vie* in sufficient quantity, but he managed to send a case with three bottles of insects that made it to their intended destination in due course. Thanking him for the shipment, Réaumur pointed out that "in the future you can very well use the distilled liquor of your country, which is apparently made from sugar." Correspondents in Saint Domingue and Martinique had already successfully used Caribbean rum, or tafia, for the purpose. To further economize, the insects could be packed more densely; each bottle could hold "as many as you can get in, without squashing them."[81] Réaumur also asked politely that the specimens be labeled with ordinary ink on little parchment tags, "attached to the body or to some other part of the little animal," noting the plants they live on.[82]

Artur was a valuable resource, especially because he lived for such a long period in Cayenne and devoted considerable attention to local flora and fauna. He observed the tropical ants and "cardboard wasps" Réaumur knew from travelers's accounts, and sent samples of the wood the wasps masticated for their paper nests.[83] Réaumur enlisted his colonist correspondent to observe the insects' behavior, their habitat, their food, and so on; ideally, Artur would become a proxy for the naturalist, who recommended "seeing for oneself" whenever possible. Once Réaumur turned seriously to birds, several years into his correspondence with Artur, he passed on his new preservation methods. The simplest approach was to dry the birds in an oven, after the bread was removed. A feather placed in the oven, if it did not burn, would indicate a safe temperature:

> There are no birds from your country, whether land or sea birds, that I would not be pleased to see, but I would like above all to have a series of the parrots found there. Instead of putting them in a pot or on a spit [i.e., instead of eating them], I would be very happy if you would agree to put them in the [drying] oven.[84]

When he first sent a copy of his printed instructions, he recommended wrapping the specimens in cloth and immersing them in alcohol.[85] He also asked for tropical quadrupeds. Compared to insects, hunting and preparing birds and quadrupeds for the transatlantic journey was both time-consuming and expensive. Though Réaumur repeatedly offered to reimburse the cost of materials, Artur seems not to have accepted this offer, which would have emphasized the status difference between the two men at a time when the colonist was trying to improve his social and political position.

Figure 6.9. Wasp nest from Cayenne, drawn from specimen in Réaumur's collection. Réaumur, *Mémoires pour servir à l'histoire des insectes*, vol. 6, plate 21. Courtesy of History and Special Collections Division, Louise M. Darling Biomedical Library, UCLA.

✵

In practice, Artur was not the ideal naturalist Réaumur imagined. He had to send versions of the instructional pamphlet to Cayenne several times, never sure if they had arrived safely because Artur seemed to have overlooked its lessons. Réaumur reiterated key points and further instructions in lengthy letters. Each time he received a shipment, he would remind his correspondent of precautions that might yield better results, going into considerable detail about the frame device for posing birds while they dried. He had taught the technique to "several ladies," he said; indeed, it "requires less time to do than I have spent on describing it."[86] Realizing that written description might not convince Artur to build one of these himself, Réaumur arranged to meet another colonist on the eve of departure for Cayenne, to show him how to build the frames and arrange the specimens. "After having received a little lesson in person, if he is willing to take with him a machine for small birds, it will be a model for others that could be made larger according to the size of the birds."[87]

Many times Artur's efforts came to nought. Some boxes were destroyed by voracious ants even before finding a place on a ship; birds were sometimes charred in overheated ovens; jars of specimens were packed with spirits too weak to preserve the fur of quadrupeds; letters arrived illegible from their soaking in seawater; and in time of war, shipments were sometimes captured by enemy captains. In 1746, a ship carrying chests of specimens and a long letter from Artur was almost safe in the harbor at Bordeaux when it was seized by English privateers. When he discovered what had happened, some months later, Réaumur thought he might have been able to retrieve the shipment from England if he had heard about it sooner. He regretted not having told Artur to mark the cases, in addition to his own name and address, a provisional secondary address: "*and in the circumstance that the vessel be seized, [send] to M. Folkes president of the Royal Society of London.*" Everything would then have arrived in London instead of Paris, and Folkes "would not have neglected to send them to me, being a very good friend of mine."[88] The next time English pirates captured a chest sent by Artur, it did have Folkes's name on the label, but in spite of this precaution it never reached its destination in Paris—although Réaumur eventually received Artur's letter with a list of the missing contents. "English (and other) corsairs," he wrote to Trembley in frustration, "are not particularly concerned with the progress of natural history. As I imagined that they cared little for captured goods like my box, I took the precaution of having it addressed to M. Folkes, but apparently they did not deign to pay him the compliment of sending it to him."[89]

In spite of the obstacles Artur encountered, a good number of insects, birds, and mammals from Cayenne and the surrounding region did make it into Réaumur's cabinets.[90] It would be impossible to chart the provenance of all the items in the collection inventory, or even to enumerate all the people who supplied specimens over the years. Réaumur knew virtually all of these people himself; just as he did not purchase specimens in shops, he did not solicit items from anyone without some sort of a personal connection. Sometimes the friend or acquaintance would negotiate with a third party, as when Trembley convinced the owner of an elephant skin in London to send it to Paris, or when Séguier negotiated with Veronese pet owners. Many, if not most, of the people who sent objects had visited the collection themselves, and some of them had collections of their own.[91] Some were aristocrats of Réaumur's standing or higher; others, like Artur, were linked by patronage connections to the capital, either through the administration of French colonies or through the Compagnie des Indes.

Natural history collections sought to extract and condense nature into the space of a few rooms; some collections were reduced further, from three dimensions to two, in images, inventories, or catalogs. Static displays also collapsed time, by fixing the continual change and movement of life in immobile objects and images. A grouping of objects—a pair of adult birds posed together with their nest and eggs—represented the dynamics of animal life and made it accessible, in stylized form, to the spectator or the naturalist. Ideally, preserved objects would serve as a window onto nature, or perhaps better, a door that would inspire the viewer to pass from the collection out into the world of living things, where there were always more observations to be made and things to be collected.

While regarding this extract or abridgment of nature in Réaumur's cabinets, visitors were also seeing another kind of extract, since the displays of objects represented in material form his years of scientific work. There were the clay samples used in experiments aimed at developing French porcelain; chunks of different kinds of iron and steel, recalling the work of the 1720s; a collection of spider webs and other insect nests that might remind viewers of the trials with spinning insect-produced fibers; and tubes of caterpillars and chrysalises floating in alcohol, which evoked the endless observations recorded in the insect volumes. Attentive visitors could also detect in the objects, and in their associated inscriptions, a map of the interpersonal and political relations that made such a project possible. Réaumur's cabinets transposed an intricate net of social relations—including friendships, patronage relations, and exchanges of favors—into arrays of objects and labels, where visitors could study and admire them. At least some of these viewers were recruited as contributors to the project in their turn.

The outlines of this social map survive in the six volumes of bird descriptions that make up Brisson's *Ornithologie*, published three years after Réaumur's death, when the collection no longer existed as such.[92] The books, lavishly illustrated with 220 plates drawn and engraved by François-Nicolas Martinet, commemorated the collection where Brisson had worked for eight years, mounting, observing, arranging, and rearranging hundreds of specimens.[93] Brisson grouped birds into orders (based on structure of the feet), genera (based on beaks), species, and varieties, composing his descriptions from individual specimens—many of which he himself had prepared from the hundreds of shipments sent to Réaumur from all corners of the globe. He listed vernacular names in as many languages as possible, gleaned from the tomes in Réaumur's library; he noted the natural geographical range of each species, and then the location where each specimen had lived. In some cases, especially for the more exotic birds, he also named the person who had sent the specimen, or who had been the intermediary: "Sent to M. de Bentinck, who made a present of it to M. de Réaumur"; "It is found in Senegal, from where it was sent to M. de Réaumur by M. Adanson"; or "One finds them in Germany and in the region of the Alps, from where it was sent to M. de Réaumur by Madame de Tencin."[94] Some names—Artur, Bentinck, Morton, Séguier—are familiar from the correspondence. Many others are only retrievable through these notations, taken from the specimen labels; most of these people, like Madame de Tencin, a close friend in Paris, would have examined or admired the collection themselves.[95] Naming contributors was a form of recognition, of course; many of the contributors would have viewed their own specimens, suitably mounted and displayed, when they came to visit. The labeling practice also lent legitimacy to the collection itself, putting on display the global reach of Réaumur's connections as well as the specimens themselves. When Brisson included the names in his book, he was memorializing the web of people and practices that had made the collection possible, as well as the scheme he had devised for arranging the hundreds of birds on the shelves.

7

Chickens, Eggs, and the Perennial Question of the Generation of Animals

CABINET, LABORATORY, AND POULTRY YARD

In his six-volume work on ornithology, Brisson based his description of the common domestic chicken on birds he would have seen every day, in the poultry yard. Other specimens described in the book—five-clawed hens and roosters, the Persian or rumpkin, the crested, the "*frisée*," the Japanese, Turkish, Paduan, and several dwarf varieties, as well as guinea hens and turkeys—had also lived in Réaumur's flock before being stuffed, mounted, and displayed upstairs in the collection.[1] The fowl pens and chicken houses were conveniently situated in the front courtyard of the house on rue de la Roquette, between Brisson's lodgings and the main house where he worked in the collections, the library, and the laboratory (see fig. 6.1).[2] The flock of domestic poultry was a living collection, a counterpart to the collections in the museum, and yet another site for the "particular study of ornithology." This science, Réaumur often said, had been sadly neglected: "We have not paid as much attention to all the uses it procures for us, and we have not sought sufficiently to extend [these uses]."[3] Brisson's descriptive and systematic work was only a first step toward a science that would ideally do for birds what Réaumur had done for insects.

Réaumur said that his bird collection had stimulated "observations, notions . . . and experiments"; he could easily have said the same of the birds living in his yard, under constant observation and always available for experiments.[4] The same people who unpacked shipments of specimens and prepared them for the cabinets, as well as Réaumur and his gardener, were often working outside in the *basse-cour*. They fed a vulture chunks of meat on a string to

Figure 7.1. F.-N. Martinet, chickens from Réaumur's collection. These specimens had lived in the poultry yard at rue de la Roquette. M. J. Brisson, *Ornithologie*, vol. 1, plate 17. Wellcome Library, London.

⁕

study the chemistry of digestion. They blocked the nostrils and ears of a rooster to see if this would affect the timing of its crowing. They plucked feathers to see the patterns of growth on different parts of the body. They coated eggs with a variety of oils and waxes to prolong their freshness; they tracked the behavior of setting hens; they took chicken and duck eggs and gave them to setting turkeys to hatch out. In the coops, they paired off chickens for experimental crosses, to test conjectures about the transmission of traits from hen and rooster to their offspring.[5] In short, while a great deal of effort went into obtaining, preserving, posing, and eventually describing and drawing the dead specimens, hundreds of chickens, turkeys, and other birds lived as experimental subjects in the courtyard.

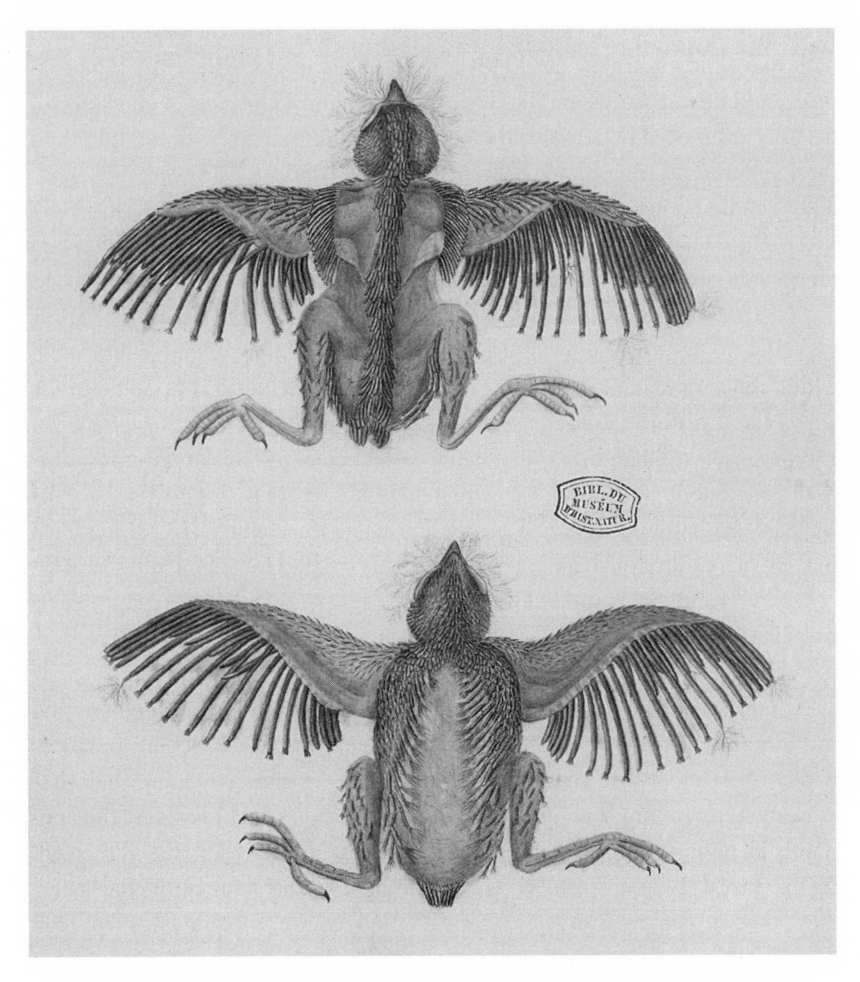

Figure 7.2. Hélène Dumoustier, partially plucked birds, for study of feather growth. Ms. 972, BCMHN. By permission of Bibliothèque centrale du Muséum nationale d'histoire naturelle, Paris.

◦

In the 1740s the experimental approach Réaumur had applied over the years to industrial processes like steel and porcelain production, to insect physiology and behavior, and to the preservation of specimens, took over the poultry yard as well.[6] In parallel with the rapid increase in the numbers of dried and stuffed birds in glass cases, the domestic fowl proliferated, as Réaumur and his gardener Guillaume developed methods of artificially incubating eggs and raising chicks in large numbers. Initially, they harnessed the heat generated in dung heaps, in which they buried barrels of eggs. Then they experimented with spaces adjoining various kinds of wood-fired ovens, where the temperature could be monitored and held constant. Réaumur's interest in chicken cultivation was rooted in both ornithology and agronomy, and he brought both of these into the meeting room of the Academy of Sciences. The academy's mandate included, after all, improving technology in the service of the kingdom's prosperity. What could be more economically beneficial than making good on Henry IV's vow to put a Sunday chicken in every peasant's pot?[7]

"The procedures, which I managed to discover only after pursuing fruitless experiments for a whole year, are nevertheless as simple as one could wish, within reach of all country people," Réaumur told Séguier. "I hope that they will make it possible to increase domestic fowl production in all regions. . . . I have reason to hope that this practice will become established because it is both convenient and useful."[8] In the meantime, whatever long-term benefits artificial incubation and cultivation might bring to the countryside, raising birds to populate poultry yards and pleasure gardens was fashionable. Ladies kept diverse flocks and experimented with incubating eggs; Réaumur's cook prepared the chickens hatched artificially for his guests, and sometimes even served up, as a special delicacy, the tiny chicks still in the shell.[9]

At the time of Réaumur's death, the notaries counted forty-three hens and roosters and thirty chicks in the coops. In the heyday of the breeding and incubation experiments ten years earlier, there were considerably more than this, as many as three hundred birds at one point.[10] Visitors could not miss the constantly changing population of birds and other animals as they passed through the courtyard on their way to the collections upstairs. Strolling by the fowl pens, they could inspect new inventions: special "lodges" for isolating pairs from the flock, incubators set into manure beds, and "artificial mothers"—heated containers designed to keep newborn chicks warm. If they were lucky, they might witness the thrilling spectacle of dozens of chicks breaking free of their shells. The living flock and the arrays of preserved specimens were all part of the natural history pursued in the yard and stables, the collection, the laboratories, the library, and the study. The assistants, some of whom lived on the premises, did a variety of tasks, though each had their own specialities. Hérissant, a licensed physician, did all manner of dissections, as well as chemical analyses for the digestion experiments and trials with preservation techniques.[11] Father François Menon, who died prematurely after only eighteen months in the household, maintained the pickled specimens in the collection, and worked with the eggs and chicks in the incubators. Brisson, as we have seen, worked on all aspects of the collection; he was known for making and testing thermometers and spent a good deal of time in the library and at the writing table, composing his systematic works on quadrupeds and birds. Each of the men developed particular ways of sealing jars and posing birds.[12] Meanwhile, Hélène Dumoustier was drawing nests, beaks, and feathers (see fig. 6.3). In his study, Réaumur answered letters, marked eggs, and made notes on his experiments, while outside the gardener monitored the temperature in the dung heaps with Brisson's thermometers.[13] The lives and habits of domestic fowl presented conundrums, problems, and opportunities for the observer intent on "perfecting" ornithology as well as increasing poultry production. Experiments on domestic fowl fit perfectly into the "curious and useful" model of experimental natural history.

ARTIFICIAL INCUBATION OF EGGS

Réaumur spent more than a year experimenting with incubating eggs before he reported successful results to the public session of the Academy of Sciences in November 1747. According to the Swiss mathematician Gabriel Cramer, who happened to be visiting Paris at the time, Réaumur regaled a crowded room of enthusiastic listeners with the story of his setbacks, frustrations, and ultimate success. Although conceptually simple, keeping the incubator at constant temperature and low humidity required vigilance and ingenuity, as well as skill. To the academy Réaumur framed this both as a scientific problem requiring exact measurement, systematic observation, and judicious reasoning, and as a technique simple enough to be used by peasant women. As Cramer remarked, "This shows that this accomplished *physicien* is always turning his mind to public utility. . . . He made everyone see how useful it would be for the kingdom to increase poultry production." In the experimental beds in rue de la Roquette, the target temperature was maintained with thermometers made on the premises. In more rural settings, peasants could achieve the same results with a homely substitute for the thermometer using butter mixed with wax or suet. "As this machine [the thermometer] is too delicate for the clumsy hands that must occupy themselves with this, [country people can use] a mixture of two parts of butter and one of beeswax, which should be the consistency of thick syrup in the barrel [i.e., the incubator]."[14]

The published account of the incubation experiments followed the same narrative strategy as the engaging talk to the academy's public audience. As he had often done in his insect books, Réaumur told a tale of discovery punctuated by triumphs and disappointments, discoveries and setbacks, always stressing his persistence in the face of failure. The long string of observations "finally opened my eyes to the outcome that I had sought in vain for such a long time."[15]

The gardener played a key role in the saga, monitoring and adjusting the temperature in the beds, and generally taking care of the eggs and the laying hens. Indeed, he became just as invested in the hatching experiments as his master, if not more so. "A hen does not love her chicks more than he loved those he had hatched [in the incubator]; when he saw them perish it was not with that indifference with which, it seems, the hen watches her own chicks perish."[16] Finally, after months of disappointment, the gardener "who had cared for so many unfortunate batches of eggs, and whose hopes had been sustained like mine, came in one evening, completely beside himself, to announce to me the news we had anticipated for so long. . . . One of the eggs was crazed, that is to say, it had little cracks in its shell and the little chick that had made them could be heard within."[17] The climax of the story was the welcome spectacle of the eggs coming to life in the baskets, "a pleasure which I enjoyed every day after that" when opening the cask to insert new eggs.[18] This was not to say that every batch was successful; it took some time for the gardener to become expert at gauging and adjusting the temperature and humidity in the beds. But from this point on, artificial incubation became practical, as well as enthralling.[19]

Initially, the incubator was little more than a box filled with eggs and placed carefully in a pit dug into a manure pile and covered with planks. Daily inspections, by breaking an egg and looking for the heartbeat of the embryo, indicated that all was well for about a week. When the embryonic chicks started dying, even at the correct temperature, a long chain of trials and observations, with the sacrifice of many eggs, ultimately pointed to moisture-laden fumes as the cause: "Finally, I remembered that the eggs in the incubator had spoiled more quickly when the inside of the lid had been the wettest; my repeated experiments were thus not in vain, in that they taught me at least that the chicks had survived much longer when they were in a less humid atmosphere."[20] The damp fumes had apparently penetrated the eggshells and suffocated the embryos. To adjust the humidity, they put layers of shallow egg baskets into wooden casks with adjustable openings in the lid to vent the fumes as necessary. A central hole accommodated a thermometer, which could slide in and out to different depths in the container without disturbing the eggs (see fig. 7.3).

The first public presentation to a packed house at the academy sparked considerable interest in elite circles in Paris, where a surprising number of people wanted to try incubating their own eggs. Long before the detailed protocols appeared in print, Réaumur and his assistants were swamped with requests for consultations:

> I can barely contain the impatience here of those, especially at the court, who want to make use of [my method] without being fully instructed in it, and above

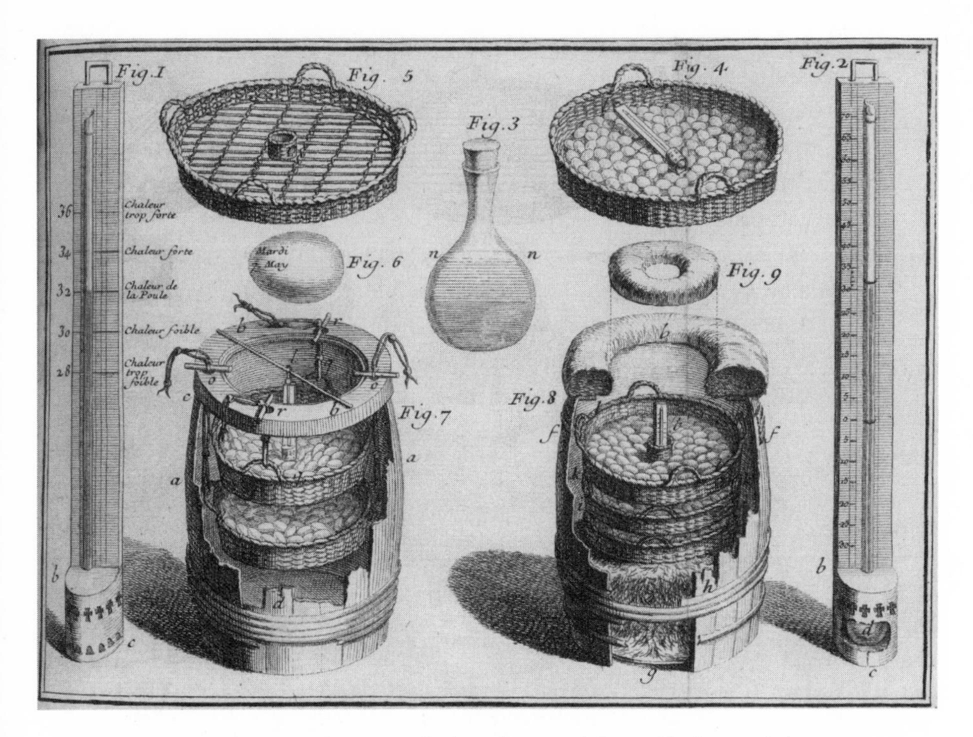

Figure 7.3. Pierre Soubeyran, hotbed incubator with layered baskets and alternatives for placing thermometers: "chicken thermometer," with key temperatures marked (fig. 1); butter thermometer (fig. 3); basket with small thermometer on eggs (fig. 4); basket with central opening for long thermometer (fig. 5); egg with date marking (fig. 6); thermometer suspended from barrel rim (fig. 7). Réaumur, *L'art de faire éclorre et d'élever en toute saison des oiseaux domestiques*, vol. 1. Wellcome Library, London.

> all by the impatience of our princes and princesses. The priest of Saint Sulpice already has an operation that is succeeding well, at his Community of the Infant Jesus. The people who run it came to take sufficient lessons from my gardener, who is the great master of this new art.[21]

In his book, a comprehensive illustrated guide to hatching and raising chicks, Réaumur described the setups devised for various locations around town, to show how the procedure could be adapted to different places and situations. He also reported on all this activity to more distant correspondents, and many of them took it up as well. In the printed text and in letters, then, we can glimpse the informal circulation of practices adapted to particular locations and circumstances by a variety of people, initially in the Paris region, but soon extending well beyond.

Although the method was intended ultimately to revolutionize practice by farmers and peasants in the countryside, many of the early enthusiasts were nobles whose experiences in their poultry yards lent a certain cachet to the otherwise humble task of raising chickens. "The wife of President Ogier wanted to be one of the first to hatch chickens by means of dung heaps," and she soon had her servants constructing the hotbed, "to try this experiment before her eyes." This lady's barnyard contained a perfectly situated manure pile, protected from the elements by an open roof attached to the stable wall; the location was even better adapted to the purpose than the open shelter in Réaumur's garden.[22] A similar structure protected the incubators established in manure beds at the new Versailles menagerie.[23] The proliferation of these "chicken ovens" meant that readers could learn from the experiments of princesses: "A great princess had reason to be dissatisfied with a bed that she had had made in a low place, small and quite poorly ventilated; the interior of this space was continually filled with a vapor thick enough to be visible, which could only be deadly for the embryos."[24] Réaumur's old protégé Nollet, in his capacity as physics lecturer to the French court, experimented with a manure hotbed at the royal pastoral retreat of La Muette. Nollet had his own quarters in the gardens at La Muette, where Madame de Pompadour, the king's mistress, had a newly remodeled dairy as well as a scientific instrument collection. In the first attempt at artificial incubation he lost two thirds of the chicks just before they were ready to hatch. The problem was evidently caused by excess moisture in the beds; Nollet experimented with different techniques and found that he could keep the eggs dry by removing the baskets of eggs from their casks several times a day to air them.[25] Réaumur told this story in print to show how the experimental habits of mind that served the physics demonstrator so well could also be put to use in the poultry yard.

Women and men of aristocratic rank took their bird collecting and poultry cultivation seriously, and many of them rode out to Faubourg Saint-Antoine to visit Réaumur's yard. The Marquise de Broglie sent her gardener: "He was intelligent, and after twenty-four hours he thought he was adept enough and left to spread this knowledge in the neighborhood where he lived." His master, the Marquis de Broglie, collected partridge eggs from nests exposed in harvested fields; the gardener successfully hatched them in the manure beds and released the partridges back to the fields—no doubt to be hunted.[26] The director of the Versailles menagerie, sparked by the king's particular interest in the poultry yard, sent a boy to Réaumur as well. After "three or four lessons," this boy knew how to incubate the eggs of many different kinds of birds. Manure beds were put into service at Versailles to hatch out pheasants, partridges, ducklings, peacocks, and guinea fowl, as well as chickens. The king himself followed their progress, witnessing "with his own eyes" the spectacle of the baby birds emerging from their eggs, and even helping free them from the shells.[27] Réaumur was invited to Versailles to admire the chickens and many other kinds of birds hatched and raised according to his instructions. As a special mark of favor, when the last surviving female ostrich in the menagerie laid a few eggs in the summer of 1748, the king ordered that one of them be sent to Réaumur's yard to be hatched artificially. (The ostrich refused to set on her own eggs.) In spite of the perfect conditions and careful monitoring, after five weeks it became apparent that the ostrich egg was not healthy, and the experiment concluded in disappointment. A baby ostrich would have been extravagant recompense for the years of experiments. At press time, Réaumur continued to hope that eggs from some newly imported female ostriches might produce chicks in the next season.[28]

Though the small manure beds yielded the first artificially hatched chicks in rue de la Roquette, Réaumur had long been intrigued by the possibility of harnessing wasted heat generated by ovens and furnaces. Travelers in the seventeenth century had marveled at the large-scale operations they observed in Egyptian villages, where tens of thousands of eggs were incubated at a time, using slow-burning fires to heat buildings dedicated to the purpose. European princes had at various times tried to replicate their methods; the Grand Duke of Tuscany had even imported an Egyptian villager who knew how to build and regulate these ovens. Réaumur opened his book with an extended description of the Egyptian stoves; his own trials using the heat from bakeries in his neighborhood effectively domesticated the mysterious foreign practice.[29] The "secret" closely guarded by the Egyptian egg experts reduced to keeping the warm room at the correct constant temperature.[30] Thermometers, he noted, were more reliable guides than the "long experience" of these men; once he had determined the temperature under a setting hen, anyone could use thermometers to monitor temperatures and adjust the heat source accordingly.

THE CHICKEN OVENS OF PARIS

Incubators using heat generated by bread ovens were installed in several locations around Paris in the winter and spring of 1748. The philanthropic priest of Saint Sulpice parish, Languet de Gergy, was one of the first to ask for advice. Languet, a well-connected aristocrat, had founded a residential community in his parish, run by nuns, with a small convent school for girls from impoverished noble families as well as a workhouse employing a hundred destitute women and girls in the production of cotton textiles. Languet had long been interested in novel technologies, and he may well have been in the audience at the Academy of Sciences when Réaumur read his paper.[31] The Community of the Infant Jesus operated a bakery and a farm on its premises in the Faubourg Saint Germain, with pigs, a dairy, and lots of poultry; the proceeds supported the parish. A thriving operation in this well-known charitable institution was exactly the kind of example that might inspire other places to try artificial incubation, and Réaumur told the story in some detail.

Languet had invited him to the community to inspect the premises and to advise on design and placement of a hotbed suitable for hatching eggs; after settling this, the priest took him on a tour of the compound, including the bakery, where two large ovens operated around the clock. Above the vaulted ovens was a sort of attic beneath the roof, accessible through a small door let into the wall just above the ovens. "I climbed up into this room; the heat I felt as soon as I entered made me suspect that it could be a stove as convenient as the Egyptian ovens for hatching chicks."[32] Some weeks later, he came back with a thermometer to confirm his suspicion. He recommended moving the incubator to the bakery, even though the nuns were already having some success with hatching chicks in the manure-fueled hotbed. Réaumur thought that the space above the oven would make the work "less trouble and . . . more appropriate to young women who like cleanliness."[33] Father Menon, Réaumur's assistant, went back the next day, armed with seven or eight thermometers, to map out the temperature variations in different parts of the small room above the bakery and to recommend the best location for the eggs. The nuns happily took charge of the first experimental batch of a few dozen eggs, equipped with Réaumur thermometers and detailed instructions, and they were rewarded with newborn chicks three weeks later.[34] The nuns continued to use both sources of heat to hatch out a steady stream of chicks. Sister Marie in particular devoted herself to the chicks, making "repeated experiments" and achieving remarkable success: 296 out of 300 eggs hatched successfully under her supervision.[35] Réaumur could hardly help boasting of this success. In the community's yards, he reported to Father Mazzoleni, "chickens swarm around everywhere even though some of them are sold every day."[36]

Back in his own neighborhood, Réaumur experimented with the oven of a baker just across the street from his house. Though the space above the oven was smaller, no more than a few inches high, it could accommodate eggs easily, and the servants learned soon enough how to shift the eggs around to compensate for the fluctuating temperatures produced by the weekly baking cycle. Another stove in the neighborhood was at the priory of Bon Secours, whose poultry yard and bakehouse backed directly onto Réaumur's garden. The Abbey of Malnoue, where Hélène Dumoustier's sister Catherine had been a nun for many years, had recently moved to this location, and the abbess, who happened to be a friend of Réaumur, agreed to assign some of the nuns to the task of supervising the incubator.[37] The nuns, equipped with "chicken thermometers," were eager and willing to learn the new technique.[38]

Once the wrinkles had been ironed out, Réaumur's poultry yard, with its hotbeds, incubators, and specialized pens of various kinds, welcomed visitors, just like the collections upstairs. The architect Blondel, in his description of the house, mentions not only the museum occupying most of the second floor, but also the yards and gardens, with large parterres, an orangerie, alleys, hedges, and a kitchen garden. "One can also see in the *basse-cours* of this house a place designed specially for hatching chicks, following M. de Réaumur's method. These curiosities are also presented to the public wishing to know something about them."[39] To Lord Morton, no longer in Paris, Réaumur wrote, "I regret that this research was not brought sooner to the degree of perfection where it is now. I can imagine that the ladies [Morton's wife and daughter] would have been drawn to visit me by the pleasure of seeing little chicks hatch in their hands, and watching them swarm around in the places where I raise them."[40] The yard must have been a lively place, with people coming and going among the hundreds of birds, including many varieties of chickens, turkeys, ducks, pheasants, guinea hens, and peacocks. Sometimes there were sights more odd than beautiful, like the sheep with two extra legs growing from its chest "like a cravat," brought back one autumn from Poitou. It lived for several years in the yard, and occasionally indoors, as a pet and a curiosity; eventually its stuffed carcass found a place in the collection.[41]

By the time Réaumur's book on chicken cultivation came off the presses, two years had passed since the audience at the academy had heard a preview of the work in progress. In the meantime, many people had visited the poultry yard in Faubourg Saint-Antoine, and some of them had experimented for themselves.[42] The book had expanded to fill two volumes, illustrated with fifteen plates and ten vignettes. It was a handy size, but rather longer than the author had originally intended. Detailed instructions for building and operating the chicken ovens alternated with observations of how birds hatch their eggs in nature and with reports on experiments, whether failures or successes. The reader learned about how a hen moves her eggs around to keep their temperature constant; how the chick develops, especially in the final stages; the technique used by chicks to peck at the shell when ready to hatch; the capacity of turkeys to set on large numbers of chicken eggs; the behavior of hens and

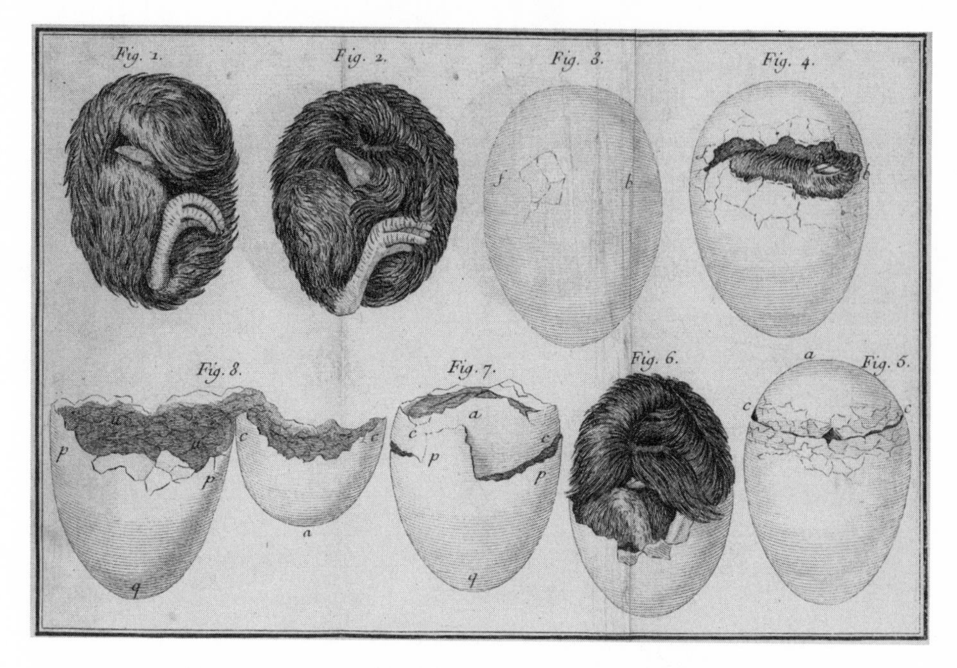

Figure 7.4. Stages of chick breaking through eggshell. Réaumur, *L'art de faire éclorre et d'élever en toute saison des oiseaux domestiques*, vol. 1. Wellcome Library, London.

capons toward chicks; and so on. The book was a natural history of domestic fowl, as well as an instruction manual for hatching and raising chicks.

The methods of chicken production attracted the attention of nobles and others with a variety of motivations. Some of these could certainly be classed as "amusement," though others were "economical" or useful. Réaumur recognized that the full impact of artificial incubation would only be felt if it reached another layer of intended recipients. If the incubation methods were to increase the nation's poultry and egg production, they would have to travel beyond the poultry yards of aristocratic ladies and nuns to those of farmers and peasants. Réaumur imagined this dissemination proceeding in paternalist fashion, via landowners and clergymen, who would instruct their tenants and parishioners after having read his books. To this end, he wrote a drastically abridged version, released in 1751 at the same time as an updated and expanded second edition of the full text; he designed the short *Practical Art of Hatching and Raising Poultry* specifically as an instruction manual, without the narrative embellishments.[43] However grounded in experimental method and expert observational practices, his examples showed that the method was well within the reach of ordinary gardeners or servants or peasant women; only the experiments and the reasoned deductions were difficult.

The full-length book addressed another set of readers, those well versed in physical sciences, who might even do further experiments:

> The responsibility for multiplying poultry, which is currently abandoned to country women, would thus be worthy of occupying natural philosophers [*physiciens*] who have a greater talent for observation, for imagining experiments, and the persistence to follow through on them: it would be desirable if they chose to spend their time perfecting what I will only sketch out here.[44]

The poultry investigation was always conducted in multiple registers: practicality and utility for use by farmers and rural households alongside amusement and enlightenment for those of "another station," some of whom might also appreciate the "philosophical" matters related to ornithology and to the vexed question of generation. These registers overlapped at times, as when the aristocrat Languet de Gergy promoted the innovation not only for its philosophical value, but because his charitable institution had many mouths to feed.

READERS AND CHICKENS

Though it would be impossible to trace systematically the effect of Réaumur's book on the practice of poultry cultivation, a great many readers did go out into their yards and gardens to build (or to have their gardeners build) hotbeds for hatching eggs and artificial mothers for raising the chicks. Some of these were friends and acquaintances who received the book as a gift from the author. Others learned of it through the periodical press and bought it from their booksellers. Still others borrowed the books from neighbors or friends and collaborated on their incubation endeavors. Many readers, including some who did not know Réaumur personally, wrote to him with their questions, problems, successes, and inventions. The volumes became part of the elaborate circulation of letters, knowledge, people, and objects that sustained Réaumur's collection and natural history more generally. Incubating and raising chicks became part of this complex of practices, reinforcing the connections linking natural history experiments, specimens destined for collections, the routine use of measuring instruments (especially thermometers), and the cultivation of domestic animals for food, market, or pleasure.

Many of the same naturalists who had spent long hours observing and experimenting with insects received copies of the book as gifts and tried their hands at hatching chicken eggs. Abraham Trembley, who happened to be living in England when the book came out, took on the task of promoting Réaumur's new work there. He presented a lengthy extract to the Royal Society in London and arranged with Martin Folkes for the whole book to be translated. Both the summary and the full translation were published in 1750 to a good deal of interest.[45] When the Earl of Stanhope built an incubator, he consulted Trembley (who had had no direct experience with raising poultry) for advice. Trembley in turn asked Réaumur about problems with Stanhope's stove and the necessary thermometers. These instruments were expensive in England, and not always as unproblematic as they were supposed to be by this time. Even the ones he ordered from Nollet in Paris seemed to be inaccurate. Trembley was trying to calibrate them himself, with obvious frustration.[46]

Bonnet, too, tried building a chicken oven at his country estate outside Geneva, but Réaumur was disconcerted to hear that it had not been a success. "However obligingly you speak of my little book on how to hatch chicks, I cannot be happy with it since it did not allow you to carry out your experiments successfully." Why, after all, should people with less demonstrable "attention and intelligence" have succeeded where Bonnet failed? A friend vacationing at the Duke of Rufec's estate not far from Paris had just reported that the first trial using casks in hotbeds in the poultry yard had yielded three hundred live chicks; the success inspired the duke to expand the operation to larger stoves heated with wood fires. Réaumur encouraged Bonnet to try again:

> I suspect that you used a cask that was too small or too shallow, and because of that it was more sensitive to variations in the heat. It must have been something of the sort since you tell me that the heat was considerably less in the center of your oven than at the edges, whereas in my ovens, once the heat reached a stable level it was higher at the center than at the circumference.[47]

Bonnet took this to heart and soon had better luck, especially with quail eggs; he went on to build his own version of a *poussinière* for raising chicks and planned to breed some distinctive varieties.[48] From seasoned experimenters like Bonnet to aristocrats on their estates, readers often began translating instructions into practice as soon as they read the book.

Another enthusiastic reader was Jean-Baptiste Ludot, in Troyes, not far from Paris. Ludot had been writing Réaumur long letters for years, mostly about insects, and had visited occasionally at rue de la Roquette. He was soon immersed in calculations and experiments on the economic implications of the new techniques for raising poultry:

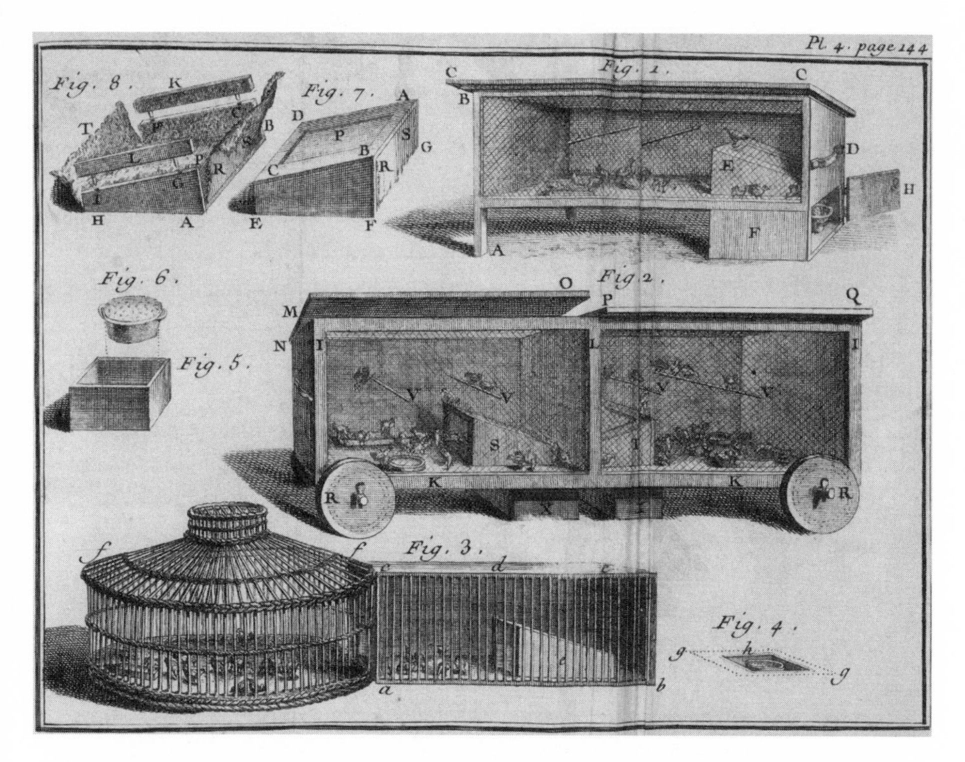

Figure 7.5. Pens for raising artificially hatched chicks: heat source (fig. 6); fur-lined "artificial mother" (figs. 7 and 8). Réaumur, *Pratique de l'art de faire éclorre en toute saison des oiseaux domestiques* (1751). Wellcome Library, London.

❖

> As soon as I received your present, I threw myself on it avidly. I read the book quickly, however, [because] I wanted to satisfy the eagerness that several people expressed for seeing it. (I have now begun to read it again.) It has circulated for a long time, and has only recently come back to me. The most philosophical of those who read it found it pleasing; it charmed those who are initiated into physical science [*la physique*].[49]

Ludot considered himself among the latter, and he did not hesitate to vigorously dispute the practicality of the large-scale cooperative village incubators, which Réaumur had proposed. Ludot promised to try his own experiments on small numbers of eggs, to test the method's feasibility for small households without enough eggs to produce dozens of chicks at a time. Ludot was an eccentric character, prone to excesses of enthusiasm and animosity, and could hardly be considered a typical reader. On the other hand, his letter shows that his small-town circle of acquaintances eagerly read new books related to natural history or technology ("the arts"), and artificial incubation must have been a lively local topic of conversation.[50]

Lord Morton wrote to Paris as soon as he received his copy of the book, to request a "chicken thermometer." This was dispatched immediately, but it broke in transit to Boulogne; Réaumur sent off two more right away. "If you are diverting yourself by hatching eggs, you may need a greater number of these thermometers; do not hesitate to ask me for as many as you would need; we make them at my house."[51] Réaumur's eagerness to supply Morton with the instruments testifies not only to the personal friendship dating to his recent stay in Paris, but also to the value of the barrels of birds he had shipped from Scotland. Because of Morton's noble status, Réaumur could not compensate him for specimens or related expenditures, so he was eager to send books and thermometers as gifts. He sent the same books and instruments to other correspondents, some as far away as Cayenne and Canada. Artur, in Cayenne, received numerous thermometers, and all editions of the book, unsolicited. Even if the busy doctor did not have the time for cultivating poultry himself, Réaumur told him, " you may have some friends who could amuse themselves usefully as people are doing at present in so many places throughout the kingdom, in foreign lands, and even in our colonies."[52]

Judging by the letters Réaumur received on the subject, a great many readers experimented with hatching, raising, and breeding chickens in the 1750s. With its combination of novelty, practicality, and systematic experimentation, the method struck a nerve with readers from many walks of life who took to heart the claim that anyone could succeed with a little care and attention: "All the knowledge that this new art demands is so simple that one can do it as soon as one has read it."[53] The "new art" also cultivated in its practitioners exactly the habits of attentive observation, including measurement and various kinds of experimentation and record-keeping that we have seen flourishing among insect enthusiasts. Like Ludot, many correspondents mentioned that they had shared the book with friends and neighbors; in some towns trials were multiplied in many different poultry yards, as readers devised variants on the basic design. Pierre Baux, in Nîmes, had three copies of the book, one in town, one in his garden, and one at his country house.[54] Some readers used the book successfully as a manual; others encountered myriad problems and wrote in frustration for advice.

A brief excursion through these letters shows that it was not always straightforward to adapt manure piles or wood fires to the task of hatching and raising chicks without hens. Most people followed Réaumur's example, however, and did not give up in the face of unhappy outcomes. When she finally worked up the courage to contact him, Madame de Laurencys, a widow in Marseilles, not personally acquainted with Réaumur, had already failed three times in spite of monitoring her stove carefully. "Your answer will not be just for me; it will be for several others in my same situation," she wrote. Like most readers asking for advice, she started with a careful description of her stove and egg boxes, with dimensions and details about the heat source and thermometers. She was using a wood fire, staying up until midnight to monitor the temperature, and turning the eggs regularly several times a day. Following instructions, she had calibrated her thermometer by taking her body temperature under the arm: the heat in the oven varied no more than two degrees from this reference point, well within the margin of safety. She opened a few eggs periodically to check on the development of the chick, and all seemed well until the very end. "The chicks were all alive, and with the yolk they occupied the whole egg from the void to the small end." She seems to have suspected her thermometer, rather than the humidity; she put it under a setting hen and got a higher reading than expected. In frustration she consulted with a knowledgeable acquaintance ("*un physicien*"), who pointed out that the scale of a Réaumur thermometer differed from the one she was using, just enough to sow confusion in her measurements. Although we do not have Réaumur's reply to this lady, he probably followed his usual habit and sent new thermometers, appropriately inscribed to show the correct temperature range and tested in his own ovens.[55]

The most frequent obstacles to incubation were excessive humidity and inaccurate thermometers; some readers who successfully hatched their eggs had trouble keeping the young chicks alive for more than a few days. Faget de Saint Martial, president in the finance bureau in Montpellier, is a telling example of a member of the provincial elite putting into practice what he had learned in person as well as in print. It so happened that he had been in Paris for the public meeting of the academy in 1747, where he heard Réaumur's lecture; subsequently he visited the poultry yard in Faubourg Saint-Antoine. Back in Montpellier, he built his own hotbed, adjusting the design several times to disperse unwanted moisture, with increasingly happy results. His first attempt had resulted in twenty-four chicks from eighty eggs; the second time he got twice as many from the same number of eggs. Still not satisfied with the yield after several years of trials, he wrote to ask for advice before making further modifications to his stove. A careful reader, he attributed the spoilage to humidity, barely detectable except by using what Réaumur called an "egg hygrometer"—with enough moisture in the air, water would condense on a cold egg placed in the bed. Faget had tried various ways of ventilating the container to rid it of vapors from the manure, to little avail. Before making a more radical modification, using shallow horizontal chests encased in lead instead of the deeper barrels, he wanted an expert opinion. "Besides making it easier to take a temperature reading, [conditions in] the chest might be maintained longer if it did not touch the dung." Réaumur had tried lining his boxes with lead for the same reason; Faget was imagining that metal on the outside might shield the contents from penetrating fumes. "Might I dare to flatter myself that you would be willing to give me a quarter hour of the time that you devote so usefully to the public, to respond as soon as possible?"[56] Evidently, the experimental impulse had traveled far beyond the immediate vicinity of the Academy of Sciences, where Faget had first been inspired to try his hand at hatching chicks.

Faget was not the only reader to devise variations on Réaumur's prescriptions. Many others adapted technique and technologies to their local circumstances and resources, and reported back to Paris even when they met with mixed results. A priest in Orléans, Picault de la Rimbertière, who had neither a manure pile nor an oven at his disposal, experimented with wine lees, the leavings from winemaking, which he knew to generate heat as they composted. Though the embryos incubating in this hotbed had developed for the first ten or twelve days, they died before they were ready to hatch. Picault speculated that the bed had been too moist and admitted that he lacked a thermometer and so had no idea of the actual temperature. "I judged that with the proper precautions, that no one is more capable to specify than you, sir, one could try this experiment and succeed."[57] Some readers tried various arrangements using kitchen fires or portable braziers to hatch relatively small numbers of eggs. One Mr. Moreau, with no available outdoor space at his residence in Brest, wanted to supplement his limited income by hatching and raising chicks in a small armoire let into the wall of a gabled room in his residence. He ordered his thermometers from Paris and sent off a detailed sketch of his design for Réaumur's approval (see fig. 7.6).

Mr. Le Duc, writing from Soissons, initially thought of using his bread and pastry ovens, but decided it would be more convenient to use the heat of his kitchen fire to hatch out his eggs. He excavated a square space in the thick wall between the kitchen and a small study, to capture the heat radiating from the back of the metal plaque lining the kitchen fireplace. He divided this warm enclosure into two levels, with a separate little door opening into each from the study, one for eggs and another for the baby chicks. He had purchased four nine-inch thermometers from Nollet, which he calibrated himself using body heat and freezing water, and used these to monitor the temperature in his stove before placing the eggs in the cupboard. "I visited these thermometers

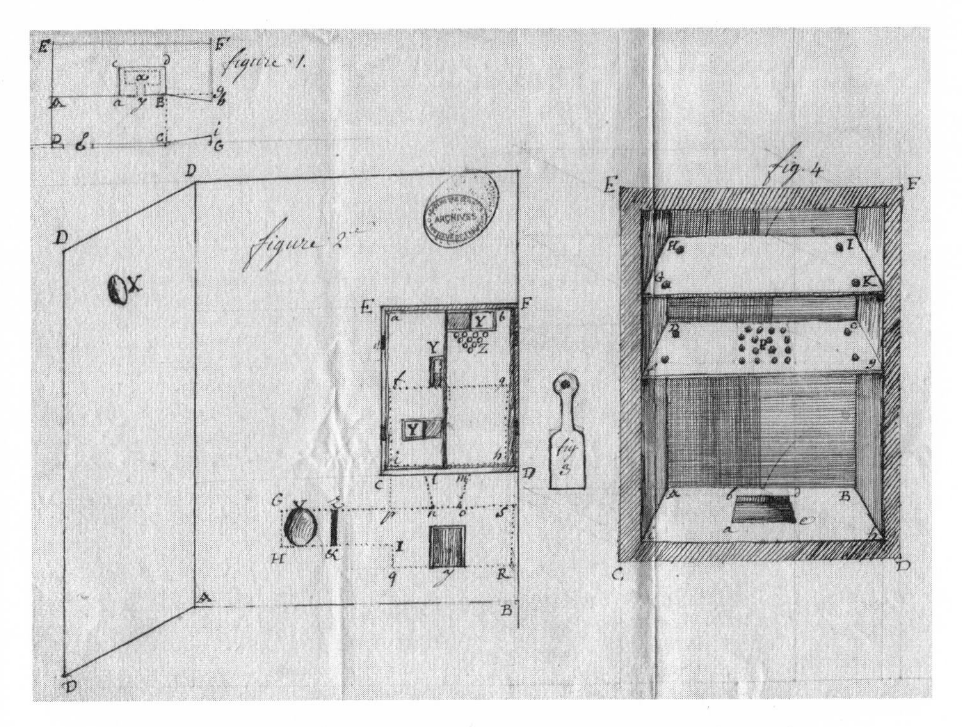

Figure 7.6. Moreau, plans of chicken stove built at his home: door to the incubator in wall of a closet (fig. 2); interior arrangement of the incubator (fig. 4). Enclosed with Moreau to Réaumur, 28 July 1752. Archives, Académie des sciences, Fonds Réaumur, Correspondance. By permission of Archives de l'Académie des sciences, Institut de France.

every hour and I wrote down precisely the degrees in each spot; I repeated this myself during two consecutive days and for the four subsequent days I gave the task to two different trustworthy people, who also wrote down the temperatures." He experimented with slow-burning coals in terra-cotta braziers to keep the nighttime temperature constant, and reported all this, with dimensions, temperatures, and so on, before any eggs had actually hatched, convinced that his design would be useful in city houses. "Without trouble, without dung, without running the risk of incommoding herself by breathing overly hot air, any lady can cut a similar oven into [the wall of] a small room behind her parlor, taking note of the temperature from time to time. It would cost her less trouble than taking care of a canary."[58]

In the pages written by these tinkerers, often equipped with the latest thermometers and books from Paris, we can see into, however briefly, a variety of experimental households, not unlike that of Réaumur—albeit on a smaller scale. Le Duc, for example, went through several batches of eggs before successfully hatching chicks, in spite of his exacting preparations and careful temperature measurements. When he tracked down the source of the problem, he discovered that his servant was allowing the oven to overheat at night. So instead of lighting the brazier inside the oven in the evening, he devised a way of insulating the eggs by placing them "between two cushions filled with a kind of thistle seed," and let his servant sleep. After three weeks, he had the first products of his efforts—eleven healthy chicks from a batch of sixty eggs. "You can well imagine the satisfaction I felt, seeing myself so rich, even though the success was not complete; [because] there was still a bit too much heat at night." Le Duc's story of his successes and setbacks, adjustments and innovations, disappointment and elation, echoed what he had read of his mentor's checkered experience. Like many other readers, he not only followed the principles and specific instructions laid out in Réaumur's book but adapted the design principles to his own circumstances, solving problems as they came up and inventing new devices or arrangements as necessary. He even admitted that his wife had provided "several useful reflections on my incubation experiments" and that she was taking care of her own batch of eggs in his oven, getting up in the night to check on them. "I assure you that she appreciates your works as much as I do, and she is reading the history of insects with pleasure."[59]

Not everyone was so happy with their results. A parish priest in Chalon-sur-Saône, Father Mugnier, sent a long account of his experiments, "with the details of the misfortunes that accompanied them and the reflections I made subsequently; perhaps the one and the other will affect you equally with pity." Mugnier had evidently studied Réaumur's book carefully but had to work with the limited available space in his home and decided to open a space in the wall above his fireplace. He had plenty of thermometers and vents to regulate the air flow in this little oven. The eggs developed well for a few days, but he had difficulty maintaining a constant temperature and the few chicks that hatched never lived long. After repeated trials and disappointments, he gave up:

> I no longer knew what to fix on; I cast my suspicions on all sides, without daring to settle on any of them. I even suspected the thermometers I was using—no doubt wrongly, [because] I had bought them myself at Abbé Nollet's house, and he himself sold them to me. Nevertheless, it is true that held under the arm for a quarter hour it rises only to 28 degrees or a little higher; I repeated the same experiment in bed and it was never above 30 degrees.[60]

Finally, he asked for an opinion on his next idea, to build a little cupboard with an iron plate on the back, to be attached to the opening in the wall where his original oven had been. "That is the best I can imagine, to have lots of space to dissipate the heat and the means to conserve it sufficiently, without incurring too much expense for the wood."[61] After Réaumur's encouraging reply, Mugnier was more optimistic. He planned to make further changes to his stove, but he refused (perhaps unwisely) the offer of a new thermometer.[62]

The poultry enthusiasts who wrote to Réaumur after reading his book varied in social station and in the resources available to them. Some, like Comte de Rochemore in Languedoc, went to extraordinary lengths and considerable expense to build their ovens. The count wrote for advice after eight months and a dozen large batches of eggs, none of which had hatched successfully. His chronicle of disappointment started with a dung heap, which was too humid, and moved on to a wood-fired stove. He modeled everything on the arrangements detailed in Réaumur's illustrations, with warm nurseries for the chicks and special coops for breeding, "all following your measurements." His freestanding room-size stove had a brazier in the middle and baskets for eggs in each corner; the baskets were suspended on pullies so they could be raised and lowered to adjust the temperature. As so often happened, the eggs started to develop but did not reach maturity. Fearing that the temperature was dropping in the night, he equipped the attic room above the stove with a bed and trapdoor so his servant could attend to the eggs:

> In order that laziness would not keep the servant from [his nighttime duties], on either side of his bed I placed two counterweights, one for raising and the other for lowering the baskets to the height he wanted, as well as an iron rod that passed down into the stove, along which the thermometer slid, which the servant raised [to check the temperature] each time an alarm sounded. According to the degree he raises or lowers the baskets and with a shovel that also goes down through the floor to the fire, he moves the coals without ever getting out of his bed.

In spite of all these precautions, the eggs still did not hatch. Rochemore could not suspect his thermometers either, as they had come directly from Réaumur's workshop (probably made by Brisson).[63] Like so many of these stories, Rochemore's has no denouement accessible to us, since only one of his letters remains in the archive.

Some of the poultry correspondents sent or promised items for the natural history collection. Moreau, who designed the small indoor stove mentioned above, promised to ask his son to send birds from India, where he worked for the Compagnie des Indes. Moreau had already sent a copy of Réaumur's printed instructions for preserving specimens to ensure that the birds would arrive in good condition. He also sent copies of the poultry book to Bengal, where he hoped his son would establish incubators. Another reader, a Dr. Bonnel from the Gévaudon region, wrote with thanks for two thermometers and a copy of the book and sent news of a live eagle he was keeping in a cage until such time as he could send the carcass to Paris for Réaumur's cabinets. He needed advice about how best to kill and prepare the large bird, and he also asked for more small chicken thermometers. The local gentry were asking him, Bonnel said, to graduate thermometers "following your method" so that they could raise their own chicks.[64] He had built his successful oven in the spacious house of a canon of the church, "a man whose rank in our cathedral does not prevent him from cultivating physics, who understands . . . the value of your useful discoveries and who also enjoys making them known in his homeland."[65]

Another correspondent, a merchant named Bourgine, wrote from Réaumur's native city of La Rochelle for advice on an ambitious venture to build large-scale incubators to raise poultry for the local market. Bourgine knew Réaumur personally and visited the poultry yard in Paris several times. A reference in his first letter to "the modest remains of a brilliant fortune reversed by the most dreadful bad luck" hints at Bourgine's straitened financial circumstances, which may have inspired his foray into raising chickens.[66] Bourgine had trouble regulating the temperature in the first ovens he built in La Rochelle, even after a visit to the famous poultry yard in Faubourg Saint-Antoine. He devoted more than a year to his trials, attempting to pinpoint the cause of his failure to get more than one or two chicks from each batch of eggs. Reluctantly, he decided that the problem lay with the thermometers, even though he had bought them from Nollet with "blind confidence." In frustration, he wrote for advice—not only for his own benefit but at the insistence of "the whole city." Demand for chickens was high in La Rochelle, where most poultry went to provision merchant and navy ships. "But how am I to help and enlighten those who would like to undertake this," he asked, "if I am blind myself?"[67] During the vacation, Réaumur and his traveling companions (Brisson and Dumoustier) met with Bourgine in La Rochelle and probably gave him new thermometers. Soon chicks were hatching successfully in La Rochelle, and Bourgine gratefully sent gifts of local fossils (snails and fish) for Réaumur's collection.[68]

Two years later Bourgine left for India, working for the East Indies Company. Réaumur may have had something to do with his obtaining this post, as Bourgine sent thanks for putting him on the road to making "an honest fortune." En route, he encountered several people interested in raising poultry and left thermometers with them. On Île de France (now Mauritius) he claimed to have established a chicken oven at the home of a "private citizen" before sailing for India. He also passed through Île Bourbon (now Réunion), where he advised J.-B.-F. Lanux, another of Réaumur's correspondents and a "zealous *curieux*," who had given up on his ovens but renewed his efforts with Bourgine's encouragement.[69] From Pondicherry, Bourgine sent insects and birds back to Paris, as well as requests for more thermometers. Eventually he ended up in Chandernagor, where once again he planned to raise poultry. He needed Brisson to send magnifying lenses and more incubation thermometers.[70] He was also busy collecting specimens to ship back to France: butterflies, fish, birds, and shells. Even these few surviving letters—there are substantial gaps in this correspondence—reveal a complex economy of circulating favors, advice, instruments, books, and specimens, over long distances and constantly shifting circumstances. When Bourgine set up his new home in Bengal, he incorporated a yard with outbuildings for raising poultry (as he had done earlier in La Rochelle) and facilities for gathering, observing, drying, preserving, and storing specimens of all sizes. Though we know little about the details of these arrangements, or about the other members of the household, the familiar model of the experimental and collecting household seems to have traveled with him to India. His shipments of barrels of specimens linked him to Réaumur's cabinet, in the manner of many other collectors.[71] But so did his efforts in the poultry yard, adapted to the local conditions in Bengal.

The poultry books traveled around the globe in the other direction as well, across the Atlantic to Canada, where the king's physician in Quebec, Jean-François Gaultier, enthusiastically proselytized for artificial incubation. Gaultier knew the Paris naturalists and botanists from his student days. As with Bourgine, his other obligations did not allow him to spend all of his time on natural history, but with the support of the colonial governor he mobilized people all over Canada to obtain and preserve animals, plants, and seeds for eventual shipment to Paris.[72] His letters neatly illustrate the way that natural history in the colonial setting mirrored the activities of naturalists in Europe: collecting, preserving, observing, and experimenting. Raising chickens fit comfortably into this matrix of practices, where the useful—whether for medical, entrepreneurial, or just nutritional purposes—and the curious reinforced each other.

Gaultier's first impulse on reading about the chicken ovens was to ask religious communities in Quebec to experiment with the different kinds of ovens, taking inspiration from the Parisian nuns who were raising chicks in their stable yards and bakehouses. "If I am fortunate enough to find someone who can successfully repeat all the experiments you enumerate, I will undertake to have eggs brought from all parts of Canada to have them hatched by this means, and I will then be able to send you, living and dead, several kinds of birds that I believe are peculiar to this country."[73] Long before the ovens were in place, Gaultier was imagining that hatching eggs could contribute to the transatlantic collecting enterprise as well as to colonial gastronomy. In the meantime, he sent barrels of fish and birds preserved in *eau de vie*, and promised many others. Gaultier served as a clearinghouse for specimens sent from other regions of Canada and arranged with hunters and fishermen to have certain animals preserved; he was an avid naturalist himself, observing the habits and migratory patterns of birds and the peculiarities of many other animals.

Réaumur sent poultry thermometers as well as books to Canada; by 1753 Gaultier could confirm that these instruments had been put to good use by a community of nuns who were "manufacturing" chickens in the winter months. "We ate some young chickens in the month of February, something that had never before been seen in Canada," he reported cheerfully.[74] The incubators produced chicks throughout the winter and into the spring: "Your discoveries are becoming useful to the whole world." Another shipment of thermometers made the Atlantic crossing the next year; by this time other convents were successfully hatching chicks. The Quebecois elite enjoyed dining on chickens in the cold season, and convalescing patients benefited from the luxurious nourishment, especially appreciated for its novelty. "This art is beginning to make progress, and I have inspired quite a few poor citizens to build brick ovens in their homes to hatch out chicks, which will earn them some money and please the appetites of prosperous people."[75]

BACKYARD BREEDING

Incubators and chick nurseries were not the only sights worthy of attention in the yards and outbuildings at Réaumur's home. Visitors to the poultry yard stepped into an outdoor laboratory, something like a cross between a demonstration farm and a curiosity cabinet, where amusing oddities and experiments in progress shared the same space. Similar spaces could be found in the various menageries and gardens at Versailles and other royal residences, at the royal botanical garden, and in any number of other aristocratic homes. Réaumur kept a variety of unusual animals under observation. One rooster, sent by a provincial priest from the barnyard of a parishioner, changed its feather color with each molt; exotic chickens were often on loan for breeding; Bazin sent some hens of a breed raised in Strasbourg. There was a duck that liked to mate with a rooster, and other captive birds, including a black stork and a vulture with clipped wings, shared the yard with the chickens and turkeys.[76] A major attraction one season was a rabbit-hen pair whose unlikely copulation had been reported to the academy; incredulous, Réaumur had obtained the pair for firsthand observation. Sequestered in a small closet directly off his ground-floor study, where the animals could be observed undisturbed, they drew "all of Paris" out to Faubourg Saint-Antoine for the spectacle. Réaumur described the unlikely scene to Father Mazzoleni in Rome:

> There was a report to the academy several months ago of a very bizarre marriage of animals, that of a rabbit with a hen. This union, so counter to nature, caused a great uproar in Paris. I was less disposed than anyone to believe in its reality; nevertheless, it became possible for me to see it and to attest to it when the rabbit and the hen were given to me. I put them where I could observe them frequently and easily, in my closet with a low window that opens onto the garden. I actually saw the rabbit make the same advances that he would have made to a female rabbit, and the hen accepted them as she would have accepted a rooster.[77]

The phenomenon seemed sufficiently extraordinary that Réaumur kept a journal for nearly two months, recording the patterns of activity and various odd behaviors of the two animals. Isolated from the yard, the animals were visible through the large window in the outside wall of their closet—not only to the naturalist but to his guests as well. [78] The amorous rabbit and hen became one of the "philosophical amusements" chronicled in the final chapter of the poultry book, where they lent an uncharacteristically risqué flavor:

> To avoid reproaching myself for having gone on so long about what happened right before my eyes between the hen and the rabbit, I need only recall the time when all of Paris wished so intensely to be informed about it, the time when I could find no one who did not want me to recount what really happened in their coupling, when the curiosity to see them brought to my house so many people of different classes.[79]

Witnesses to the spectacle in the little closet no doubt speculated about the possible offspring of this unorthodox union, though Réaumur did not expect the hen's eggs to come to anything. "I did not anticipate seeing a rabbit covered with feathers or a chick with large ears and covered with hair. Nevertheless, I had several eggs incubated from the hen who had lent herself to such untoward intercourse. All were infertile, as one would expect."[80]

The lascivious rabbit's attentions to the hen made for an amusing diversion, the sort of barnyard curiosity that fueled witty banter. A rather different set of breeding experiments took up considerably more time in the study, where they were planned, and in the poultry yard, where they were carried out. These were controlled crosses of different varieties designed with two purposes in mind: on the one hand, to preserve the rarer varieties by preventing interbreeding and, on the other, to produce crosses of the different kinds, to see how traits were transmitted from parent to offspring. The "degeneration" of some varieties—the loss of distinctive features through indiscriminate mixing with other types—was a recognized phenomenon in animal husbandry at the time. On the estates in his own region, Réaumur had noticed that five-clawed chickens, reputed to have been introduced to Poitou from India by a ship's officer, had gone from being relatively common to quite rare due to this kind of degeneration.[81] Any poultry fancier who had taken the trouble to amass a flock of appealing variety and beauty would be well advised to prevent the "ill-assorted unions" that could lead to the disappearance of the most striking traits. Without this precaution, Réaumur warned his readers, "we would no longer see [in the yard] the species distinctive for the elegance or the size of their crests, or those distinctive for their prodigious size, or their long legs . . . or those remarkable for their extremely small bodies."[82]

In order to exploit the flock's diversity for the study of mixed crosses, Réaumur recommended the kind of long chicken coop his gardeners had built against his garden wall, with as many subdivisions as necessary to sequester the different varieties from each other and from the general population in the yard. His own yard had four of these coops, with daytime and nighttime enclosures and roosts for each set of birds, useful both for preserving varieties and for controlling crosses, by confining one rooster with a hen of his own kind and one or two other hens.

Réaumur made only passing reference to his work on crosses and breeding in his chapter on the philosophical amusements of the poultry yard; he never presented publicly any results from these experiments. On jumbled sheets of notes, though, he recorded various strategies for answering questions about fertilization, inheritance, and the origin of the germ. One page from early in the 1747 season starts like this: "25 March: I have begun to put hens into cages (1) to enable preservation of the varieties, (2) to do experiments on different kinds combined together, (3) to discover how long after copulation fertile eggs can be laid." The first coop, for example, housed a rooster with five claws, a hen of the same type, and a common (four-clawed) hen. In addition to hatching the eggs of the five-clawed hen to preserve the unusual strain, he planned to compare the chicks from eggs laid by the ordinary hen, to see if and when they might show some evidence of the five-clawed rooster's influence. In another cage, he put a black rooster and a black hen, "to conserve the type," along with

Figure 7.7. P. Soubeyran, coops for controlled breeding of chickens. Réaumur, *L'art de faire éclorre et d'élever en toute saison des oiseaux domestiques*, vol. 2. Typ 715.49.734, Houghton Library, Harvard University.

✻

a *poule frisée*, just to see what the chicks would look like. Eggs taken from these coops were marked, recorded, and placed either in the hotbed incubator or under a setting hen or turkey.[83]

Like artificial incubation, tracking chicken crosses turned out to be more difficult in practice than in concept; results were often equivocal at best. After months of trying to produce various kinds of crosses, Réaumur reflected, "My experiments on the mating of different kinds of hens and roosters are not proliferating as much as I might have hoped." The black rooster was killed by the hens after a month of cohabitation; rats ate some of the eggs and baby chicks; some eggs were infertile; setting hens sometimes broke "interesting" eggs before they could develop. Some hens mistrusted the written notations on the eggs, and pecked at them until they cracked. Even when the chicks did hatch, often "the eggshell of the hatching chick is broken into so many pieces either by the chick or the mother that you can no longer read what was written on it"—making the parentage uncertain. On top of that, with two or three hens in each cage, it was often hard to know which hen had laid which egg, and the barriers separating the birds in adjoining coops were not always secure. In short, Réaumur's efforts at systematic recording fell prey to contingencies, ambiguities, and interruptions. He seems never to have settled on a consistent way of noting down his plans, actions, and results, so his notes mix practical questions, numbers of eggs from different hens, comments on the health or behavior of individual birds, proposals for future experiments, and ad hoc explanations for problematic or unexpected results. Sometimes he went back and added information about outcomes, but answers to pressing questions are often missing from the loose sheets of notes, leaving plenty of mysteries unresolved.

On the vexed matter of generation, Réaumur kept to the safest path, which seemed to him the self-evident one: the germs of all future living beings must have been made at the creation of the world. However difficult it may be to imagine the indefinitely small germs encased in each other, to Réaumur and most of his contemporaries this model seemed more comprehensible than any of the alternatives, which relied either on Lucretian-style chance combinations of material particles or on nonmechanical, and therefore occult, forces. As Réaumur insisted, any theory based on the mixing of seminal or "prolific" fluids, with particles contributed by both parents, had to solve the problem of order: mixtures of seminal fluids without preexisting forms of some kind could only be chaotic and random:

> What agent is to going to disentangle this chaos, to sort the various parts which belong together, to form organs with them, to unite the different organs to those to which they ought to connect, in short, to finish that germ which is no less admirable for being so small that the best microscopes cannot make it visible to our eyes? We must not expect . . . that the simple action of a gentle heat could ever produce such a work, a work infinitely more complicated than any repeating watch.

The "simple action of a gentle heat" refers to Descartes's hypothetical model of the formation of the fetus, which he had likened to fermentation; for Réaumur, the more recent "fashion" for Newtonian attraction only made things worse:

> How could attractions give to one mass the form and structure of a heart, to another that of a stomach, to another that of an eye, to another that of an ear? . . . It is only too obvious that in order to arrive at the formation of such a complex edifice, it is not enough to multiply and vary at will the laws of attraction, and that we would have to attribute extensive knowledge to that attraction.[84]

In other words, attractive forces arranging bits of organic matter into organs and then into whole creatures could not be blindly mechanical; they would have to have some sort of intelligence to guide them—the old problem of design and teleology. This short discussion is the only passage in Réaumur's published work where he engages the controversial, and very recent, theories of Maupertuis, and possibly Buffon.[85]

The controversial attacks on the preexistence of germs by Maupertuis, Buffon, and John Turberville Needham made Réaumur's breeding experiments in the 1740s eminently topical. Réaumur assumed that the fully organized germ must exist before conception; his working notes show that he leaned toward locating the germ in the male, though he held open the possibility that he might find otherwise.[86] He designed experiments to test this, or to confirm it, by following the inheritance of distinctive traits in his chickens, experiments strikingly similar to those done by Maupertuis with dogs around the same time.[87] The results from Réaumur's chickens, though, were often equivocal; some cases suggested that both male and female parents contributed traits to the offspring, an essential piece of Maupertuis's theory. Indeed, though he remained convinced of the integrity of the preexisting germ, Réaumur knew, as we have seen, that indiscriminate mixing of interfertile varieties could result in the disappearance of distinctive traits, or even the appearance of new ones, over several generations. Why should mixing cause degeneration if germs were already completely formed? Réaumur never addressed this question explicitly—and indeed, it would not be settled definitively for well over a hundred years.

The five-clawed chickens promised to be useful for experiments on inheritance, because the extra claw was readily visible in new chicks, unlike some other traits. The first chicks hatched from eggs fertilized by the five-clawed rooster exhibited the extra claw, as expected, but after that things did not remain so clear-cut. Here and there, Réaumur noted his problematic findings: "I found a chick, crushed while emerging from its shell, that baffled me with respect to the preceding experiments. The shell is broken. I found only small fragments, but the notations that I think I can decipher and the size of the chick seem to prove that the egg was from the first cage from the Paduan hen [and the five-clawed rooster]. Nevertheless, it has only four claws and a rump." (The Paduan hen had no rump; thus, the chick shared one trait with the hen and one with the rooster.) A few days later, a chick from an egg in the third cage, where the rooster had no rump, seemed to have a rudimentary rump; the chick died, however, before this could be confirmed. "This fact, if it were certain, would be very troubling." Another chick from the same cage had no visible rump, like the father—though it was hard to confirm the existence of a rump in young chicks. This particular chick was fitted out with a blue stocking on one leg so it could be identified in the yard as it grew.[88]

In the second year of the crosses, things got even more confusing when a chick from a common rooster and a five-clawed hen hatched in the hotbed. This chick "undermines absolutely any system where the embryo is due entirely either to the male or the female."[89] On one foot it had only four claws, "conforming to the observations of last year, which seemed to prove that chicks have all the parts of the male and only the parts of the male." But the other foot had an extra rudimentary claw of sorts: the back claw was double. It remained to observe this claw as the chick grew. "Whatever the case, this chick resembles the male in one of its feet and the female in the other. Therefore, they both contributed to the production of the germ. Here is a very troubling result."[90]

This kind of distressing evidence, which cropped up from time to time, only spurred Réaumur to devise experiments that might give less equivocal results. Shifting the occupants of the coops so that an ordinary rooster could mate with one five-clawed and one rumpless hen, he predicted that the new arrangement should result in eggs that would give decisive answers. "If the eggs of the hen with no rump yield chicks with a rump, it will be due to the male, and if the five-clawed hen gives eggs whose chicks have only four claws, the decision will be complete." The outcome was hardly so conclusive. In due course, he broke open one of the five-clawed hen's eggs:

> The chick had died in the shell, but what absolutely throws into disarray all my preceding observations is that the chick had five well-formed claws, very distinct, on each foot. This hen had no communication with a five-clawed rooster. What to conclude from this experiment?[91]

Ten days later he opened another egg from the same hen, which likewise held a chick with five perfectly evident claws on each foot. "I cannot suspect any error, nevertheless the fact must be repeated." Similar things happened on other occasions. Chicks from one hen paired with the five-clawed rooster all had five claws, as expected, until the day when two chicks from the same parents hatched with only four claws. Could a different rooster have gotten into the cage? Could the hen have escaped to the yard for a time, without anyone telling the master? For the historian, the most tantalizing aspect of this record of activity is that Réaumur repeatedly planned crosses designed to settle the question of whether chicks necessarily resemble the father, but the results are nowhere to be found. An unexpected contingency might have interfered, when the hen broke the egg or the marking became illegible; sometimes the notes are simply missing. There is no evidence that Réaumur actively suppressed or altered his findings, but the occasional results that might have led to his questioning the received wisdom about preexisting germs did not lead there at all.

Although his unpublished notes show that Réaumur bent over backward to maintain his commitment to preexisting germs in the male (at least for birds), in print he held to a strictly agnostic position. In the final chapter of the poultry book, where he encouraged readers to experiment in their own flocks, he mentioned the two varieties living in his pens that were particularly suited for such crosses, the five-clawed and the rumpless, one with a superfluity and the other with a deficiency. He framed these crosses as experiments still to be done:

> Let us put common hens with a five-clawed rooster and five-clawed hens with a common rooster; let us put common hens with a rumpless rooster and rumpless hens with a common rooster: if chicks are born from such matings . . . it seems that we should expect them to provide facts that will decide the question. . . . The chicks should show us by the parts they have, or by the lack of certain parts, whether the germ originally belonged to the female or to the male.[92]

He went on to explain at some length the logic of this example, to inspire his readers to try the experiment—but he never said a word about his own results, or about the myriad problems he had encountered in attempting to settle this question. In light of the elaborate, if inconclusive, breeding experiments we have explored through the manuscript notes, the presentation of the crosses as hypothetical seems disingenuous at best. Some contemporary readers recognized this. Maupertuis, for one, found the silence annoying, no doubt suspecting that the chickens would have corroborated his own reflections on bilateral inheritance of traits: "I am surprised that this able naturalist, who has undoubtedly done these experiments, would not inform us of the result."[93]

Some of Réaumur's readers did try their own crosses. Jean-Pierre Pignon, inspector-general of commerce in Marseilles, started by building stoves to hatch chicks, and then collected different varieties for experiments that might determine "whether germs exist in the male or in the female or in both." He had acquired some rumpless chickens from Syria, though the roosters were cooked for dinner before they made it to his poultry yard; the hens, crossed with ordinary roosters, produced rumpless chicks, but only females. "I am waiting for rumpless roosters. I will give them local hens, and this union combined with the preceding will serve to discover whether each sex contains its own germs."[94] Another correspondent, Thomas Antelmi, living in an isolated spot in the south of France, reported on the accidental crossing of five-clawed Persian hens with ordinary roosters on a nearby estate. "Here is one of your experiments," he remarked. Antelmi admitted to a prejudice in favor of locating germs in male semen, citing Nicolaas Hartsoeker's microscopical observations of sperm, but he jumped at the opportunity to investigate the matter himself. "Your works have inspired in me a predilection for these experiments," he wrote to Réaumur. He obtained an unusual rooster from his neighbor's brood of Persian crosses, with five claws on one foot and four on the other. The four-clawed foot looked just like that of an ordinary chicken, with a pointed rear claw, so the rooster looked like a hybrid of the two kinds. The oddity prompted Antelmi to reflect on the source of the germ:

> Did the rooster or the hen contribute the germ of this chick? Or is the germ so simple that we can conceive that the male and the female each furnished its own part? Such a mechanism would be more difficult to explain; but do we not see that the mule is neither an ass nor a horse, and that it resembles the one and the other?[95]

He then outlined a whole series of crosses to test these ideas, going so far as to consider the possibility that such crosses might not give constant results. Some offspring of a given pair might have the traits of the mother and some might resemble the father, and some might have a bit of both, like the rooster with five claws on one foot and four on the other. "If these different chicks exhibit these three outcomes, what keeps us from accepting all three positions and saying that the germ sometimes derives from the male and sometimes from the female, and sometimes from both?"[96] Antelmi planned to set up an incubator in a hotbed as soon as he had the time. Like so many of these vignettes, his story does not have a proper ending, because the record breaks off here. Even so, it clearly shows that readers, even those far outside the learned circles around the academy, were well aware of contemporary debates about generation and that they took up the challenge of doing their own experiments. Antelmi, a provincial gentleman, applied the lessons of Réaumur's book to local circumstances, taking advantage of the unusual rooster and speculating about possible outcomes.[97]

BUFFON AND GENERATION

In the spring of 1748, Réaumur was starting his second season of breeding poultry. Across the Seine at the Jardin du roi, Buffon embarked on a series of microscopic observations of "seminal fluids," looking for evidence of what he would later call "living organic particles."[98] With several associates corroborating his observations and a surgeon to perform dissections, he examined fluid taken from the reproductive organs of both male and female animals. Viewed through the microscope, material from the females teemed with what he initially called "moving bodies" that looked remarkably like the "spermatic animals" in male semen. From there it was a short step to the claim that this female semen, when mixed with that of the male, contributed to the formation of their offspring.[99] By May, Buffon had incorporated his findings into a sketch of a new theory of generation, and some months later he read a paper to the academy reporting his "discovery" of female seminal fluid in dogs, rabbits, sheep, and cows.[100] Eminently pleased with the results, Buffon reported to his friend Cramer in Geneva:

> This paper was extremely well received, and it settles a great question: it is now quite certain that there are no eggs in female viviparous animals, and that on the contrary they have an animated seminal fluid which contains, like that of the male, moving bodies. [These are] spermatic animals that in every species are the same in shape and motion as the spermatic animals contained in the fluid of the male.[101]

Buffon was using "spermatic animals" colloquially here; in the formal presentation he was careful to call them "moving bodies," since he was at pains to deny that they were animals with life of their own.

To his assembled colleagues, Buffon described the bodies he had seen through his microscope but said not a word about the theory they supported. For this the academy had to wait, with the wider public, for another year, until the publication of the first three volumes of *Histoire naturelle, générale et particulière*. This ambitious and soon-to-be-controversial work, coauthored with Daubenton, was nominally structured around a description of the collection at the Jardin du roi, its scope extending to all three realms of nature.[102] Réaumur, not surprisingly, was deeply skeptical of the whole project, even before he had read the book. To Séguier, in Verona, he wrote:

> I do not know how they will carry it off, because I have seen nothing of this kind by either of them. I know that they have had extracts made of many works by naturalists and travelers, but I do not know that they have observed for themselves. The king's cabinet is not rich in insects, in minerals, or in birds; the collection of birds consists of sixty or eighty [specimens] that they had prepared in Strasbourg and which were mostly eaten by worms last year, because they did not know how to preserve them.[103]

Buffon and Daubenton's massive project was bound to annoy Réaumur, since they were moving in on his territory—especially by promising a volume on birds just when he was planning his own ornithology book and expanding his bird collection.[104] How could the inadequate collection at the Jardin du roi support a work purporting to address all of nature? But it was not just a matter of the ambitious, perhaps even arrogant, scale of the project. With the minister Maurepas's backing, Buffon was brazenly using his institutional position at the royal collection as a kind of bully pulpit for preaching about the proper method of natural history. When he did read the first volumes, Réaumur found plenty to dislike—sweeping conjectures about the formation of the earth and solar system, a theory of generation based on active matter with the capacity to organize itself, and a human-centered view of nature that implicitly bypassed divine design.

Réaumur never publicly engaged with Buffon's *Histoire naturelle*, though his antipathy to the work and its author was well known to his contemporaries.[105] He considered inserting a harsh critique into his long-promised handbook on forming and maintaining collections—but this remains in manuscript, and fragmentary. His visceral rejection of Buffon's approach to natural history comes through in this draft, as in this passage challenging the organization of the work, which ranked the animal world hierarchically, with man as the most "perfect" and the other animals ranked according to their relations with humans:[106]

> I will say nothing against a method that has revolted everyone who studies natural history. There is no reason to fear, in spite of the way it is written, that it will seduce those beginning to apply themselves to this science. They will want definite and invariable characters, that teach them where each being should be placed. And if this order, as this [new] method has it, were arranged according to the relations they have with us, everything would be arranged in a manner far too bizarre. We do not yet know whether the author would place the animals that are harmful to us far from the animals useful to us. But we see at least that wild dogs and cats will not be put in the same class as domestic dogs and cats, that the animal that is most useful to the inhabitants of one country, and which consequently deserves the top rank in there, will be much farther [from man] in another country. The Canadian naturalist will put the beaver in the first place. In China the pig will be placed before the cow. In each part of Europe, there will have to be different orders [of animals] in different regions.[107]

Viewed from the perspective of a working naturalist and collector—we might say, from a practical point of view—this organizing scheme made little sense, and elegant style ("the way it is written") could hardly redeem it in Réaumur's eyes.[108]

However successful with the fashionable literary public, however admirable his style, many contemporary naturalists agreed that Buffon could not be trusted as an observer.[109] To be sure, most of the *Histoire naturelle* did not rely on Buffon's own observations. The striking exception was a series of experiments with semen and organic infusions, liberally illustrated with microscopic views. These chapters, an expanded version of the academic paper on female seminal fluids, appeared in the middle of the "general history of animals," just after a conjectural theory of sexual reproduction framed as an explicit challenge to the general consensus about the preexistence of germs. Buffon's theory posited organic molecules sent to the seminal fluids of male and female from all parts of the body, and "penetrating forces" acting on these as the fluids mixed to bring them together into a rudimentary organized being, a "sketch" of the offspring. This embryonic being and the dynamic mixing and combination of the organic particles that formed it, lay beyond the reach of human senses, even when aided by magnifying lenses. The empirical observations of moving organic particles in both sexes of numerous species were arrayed as the supporting evidence for this frankly conjectural theory of generation: "I became more and more confirmed in the opinion that these moving bodies were not true animals, but only living organic particles; I became convinced that these parts exist not only in the seminal fluid of the two sexes, but even in the flesh of animals and in the seeds of plants."[110] Inspired and aided by John Turberville Needham, an accomplished microscopist, Buffon moved from seminal fluids taken from human cadavers and a long series of quadrupeds and fish and on to all kinds of organic infusions.[111] In these infusions, made from meat or seeds or leaves and kept in glass bottles corked to keep out the air, Needham and Buffon saw microscopic objects, "moving globules" that changed shape and size over time, and gave every appearance of being alive, or at least self-moving. Here, plainly visible through the microscope, were the organic particles Buffon had sought; they seemed to be common to all living bodies, and therefore served nicely as the organic components "seeking to organize themselves" into embryonic animals.[112] Both Buffon and Needham, in slightly different ways, saw in matter the capacity to organize itself and come to life.[113] Réaumur and many others read these interpretations as a revival of spontaneous generation in the guise of experimental science; hence the forcefulness of their objections.

When Trembley read Buffon's first three volumes, he found the reliance on bold conjecture difficult to countenance. He saw the whole project as an implicit denigration of the intensive attention he had lavished on the polyp, and indeed of empirical observation more generally:

> These volumes contain some curious facts; but not enough in proportion to their size. There are a great many conjectures, several of which are very bold. It often happens that M. de Buffon offers them first as conjectures, and then uses them as demonstrated principles. . . . [He] claims to explain almost everything about generation with a rash hypothesis. . . . It seems that sometimes he lets himself be carried away by his imagination. If his work is much appreciated, I fear that it will do a disservice to natural history by bringing back the taste for hypotheses.[114]

Trembley had always insisted on the naturalist's obligation to avoid speculation. The surprising and even marvelous things he had seen in pond water were widely reproduced by all sorts of observers, whereas no one could agree about what Buffon and Needham had seen in their infusions.

Academic protocol, if nothing else, kept Réaumur from openly attacking Buffon, whether on empirical, theoretical, or methodological grounds. Instead, he encouraged his close friend Joseph-Adrien Lelarge de Lignac to put pen to paper for a detailed critique of the first volumes of *Histoire naturelle* (in a second edition, Lignac attacked Needham as well). Lignac, an Oratorian priest with Jansenist tendencies, had been observing insects for years. He vacationed for several years at the estate in Poitou, where he joined in collecting, experimenting, and making microscopic observations.[115] In 1751 and 1752, the assembled company spent weeks observing animalcules in organic infusions, trying to see what Buffon and Needham had seen. Brisson, an able microscopist, lent his youthful eyes to the project. As Réaumur reported to Mazzoleni in 1751:

> Every day we amused ourselves upon returning from our walk, during two hours before supper, by studying the microscopic animals that appear in the water where different materials have infused. We realized that the strange consequences that M. de Buffon and Father Needham have drawn from these insects . . . are based on poorly done or false observations, and can be destroyed by observations that are not at all equivocal.[116]

In other words, Buffon and Needham were simply not reliable observers.[117] Not only did they report phenomena that could not be "seen and seen again," but, Réaumur told Bonnet, "they sought to adjust their observations to their system."[118]

Lignac's anonymous book, *Lettres à un américain* (1751), laid out in rambling and verbose prose his objections to everything from Buffon's misleading title and grandiose style to the management of the royal collection, the dangerously materialist theory of generation, and the theologically suspect account of the origins of the solar system. Réaumur facilitated the distribution, and possibly even the printing, of the book, sending it to all of his correspondents with enthusiastic testimonials for book and author.[119] Most of them appreciated it, though sometimes with reservations. Bonnet judged it "sometimes a bit too theological," though he found the style clear and precise. "The glass palace where the Academician [Buffon] lived is reduced to dust. His chimerical ideas on generation are refuted step by step."[120] Bonnet did not like the notion of self-organizing organic molecules any more than Réaumur or Lignac did. Lignac's refutation on this point rested on the claim, supported by the observations of infusions he had done with Réaumur, that Buffon's active particles were actually microscopic animals with distinctive traits, whose transformations and reproduction could be followed with the aid of the instrument. If they were indeed animals, they could hardly be the organic building blocks of all living things, as Buffon had argued.

Réaumur pushed Lignac to publish their microscopic observations of infusions, but the details of their months of work remain, like the chicken crosses, in manuscript notes.[121] Neither Réaumur nor Lignac was in a position to undertake a full natural history of the microscopic realm; their vacation observations were directed at undermining Buffon's credibility, and consequently his theory of generation. Although they did not pursue the effort, the observations at least gestured toward a natural history of the microscopic realm that would characterize different species and map their life cycles, domesticating them much as Trembley had domesticated the polyp, by dint of tireless experimentation.[122] One element of this move was a rebuff not only to the general "taste for hypotheses" that these naturalists found in the writings of Buffon and Needham, but to spontaneous generation in particular.

EPILOGUE

Réaumur's Legacy

For years, Réaumur used funds from a royal pension, awarded in 1722 for his work on steel production, for expenses associated with his collection, including stipends for his assistants. The terms of this pension dictated that it revert to the Academy of Sciences upon his death. He imagined, with what appears in retrospect to be rather naive optimism, that the institution would continue to maintain and expand his collection using this fund. He had even drawn up a summary of expenses, hoping perhaps to get a commitment to his plan from the king, or at least the minister. Though nothing came of this, the document affords a rare glimpse of the annual costs associated with such an understaking, and of its owner's vision for its continuance as a rival to the royal collection at the Jardin du roi. The undated draft, probably from the early 1750s, lists the rent of the house, stipends for a "savant in charge of preparations," plus an assistant, shipping fees for specimens arriving by post, and the cost of jars, alcohol, and other materials. This document suggests that Réaumur hoped that his collections might stay in their current location after his death, under the supervision of the academy.[1]

Réaumur died in 1757, at the age of seventy-four; in his will he left his natural history collection and his papers to the Academy of Sciences and requested that Nollet go through the manuscripts to see if anything could be salvaged for publication. His papers contained miscellaneous draft manuscripts in various stages of completion dating back to the 1720s, many of them related to the languishing *Description des arts et métiers*.[2] There were portfolios of notes, including the records of the experiments on frogs, chickens, and infusions discussed above, and many sheets of drawings, some attached to detailed observational notes, of all manner of insects and birds. There were also fair copies of nearly completed treatises on ants and beetles, five draft chapters of a book on ornithology, and most of another work on making and preserving collections.[3] Then there were sheaves of correspondance. This material was boxed up and taken to the academy, where a good deal of it remains. Like most collectors, Réaumur had given considerable thought to the fate of his museum, the fruit of so many years of work and pleasure. He surely knew that his great rival at the Jardin du roi would be interested in the contents of his display cases.

When the notaries drew up the inventory of Réaumur's collection, a commission sent by the academy was in attendance, including Nollet, Alexis Clairaut, and Buffon. With the inventory complete, Buffon apparently convinced the Count of Saint-Florentin, the minister responsible for both the academy and the botanical garden, to short-circuit the process under way at the academy. A royal decree of January 1758 put Buffon in charge of the whole operation and ordered the removal of all natural history specimens from the Faubourg Saint-Antoine directly to the Jardin du roi. Réaumur's papers were to go to the academy, though some of these ended up at the botanical garden as well, and some were kept by individual academicians.[4] As we have seen, the bird collection was partially documented in Brisson's *Ornithology* in 1760; many of the illustrations must have been made after the collection moved.[5] For his part, Buffon folded the contents of Réaumur's cases into the royal collection, and eventually incorporated descriptions of the birds sent by Réaumur's correspondents and mounted by Brisson and Hérissant into the nine volumes of the *Histoire naturelle* devoted to birds.[6] By the time these books started coming off the press in 1770, Réaumur had been effaced from the ornithological project.

By some measures, Buffon won out in his battle with Réaumur for the terrain of natural history, especially in France. Certainly he succeeded in diffusing and co-opting the material legacy of his opponent. Buffon's philosophical view of natural history was taken up enthusiastically by Diderot and made its way into numerous articles in the *Encyclopédie*, participating in the new orthodoxy of enlightenment trumpeted by that quintessentially enlightened enterprise.[7] At midcentury, it is easy enough to find expressions of the inadequacy of narrow (sometimes represented as narrow-minded) focus on particulars and of the need for general laws to put the science of life on the same footing as Newtonian mechanics. Maupertuis surely had Réaumur in mind when he called for a synthetic natural history that would subsume proliferating details under general laws:

> All our treatises on animals, even the most methodical, make nothing more than scenes [*tableaux*] pleasing to the eye. To turn natural history into a true science

> we would have to apply ourselves to researches that would allow us to know, not the particular figure of such and such an animal, but the general processes of nature in her production and preservation.[8]

The pleasure of regarding nature is not enough, on this view; all the meticulous observations and experiments behind exact descriptions must feed into a more general understanding of how nature works, if natural history is to produce real knowledge from the mass of particulars observed by naturalists.

Maupertuis articulated his frustration with the sheer quantity of observations accumulated by naturalists who never got to the big questions, such as how traits can be inherited or how matter can be organized (or organize itself) into living creatures. "What would posterity think of us," Diderot asked in 1754, "if we were able to leave it no more than an incomplete insectology, a vast history of microscopical animals? Great objects to great talents; small objects to small talents."[9] Diderot's attempt to transmit his own vision to posterity can only be deemed a success, given his central place in the historiography of the Enlightenment. Rather like Buffon, he was happy to shift his focus beyond the range of any microscope, and to turn his gaze backward into the recesses of time. "When we consider the animal kingdom, . . . wouldn't we willingly believe that there never has been more than one original animal, prototype of all animals, in which nature only lengthened, shortened, transformed, multiplied, or obliterated certain organs?"[10] The naturalist cannot make such a leap, a "philosophical conjecture" in Diderot's words, without recognizing a role for the imagination, as well as the powers of nature.

Such claims were far from the daily practice of naturalists who concentrated on what they could "see and see again." The grand views of Buffon's preliminary discourse and Diderot's *Pensées* found numerous readers, to be sure, but meanwhile, many people (even some readers of Buffon) continued to follow the precepts of an observationally grounded natural history, exemplified in the works of Réaumur. Our excursion into the world of Réaumur and his correspondents and collaborators has taken us into the gardens and studies and libraries and workrooms of many kinds of people, all intent on confirming what they had read about and on discovering anything else that crossed their paths. As Nollet noted in his inaugural lecture as professor of experimental physics:

> The *physicien* finds objects for his research and his amusement everywhere: the country and the city, the elements, the seasons, things that breathe, things that vegetate, things that are born, things that die, etc., everything offers him subjects for meditation, instruction, and profit.[11]

Nollet quite explicitly included the study of living things in the scope of physics. If, Nollet suggested, the science of physics is defined by its methods—observation plus experiment—natural history can fit into this category as well as the experimental physics of electricity or the strength of materials or the motion of billiard balls. In a sense, the experimental physics retailed by Nollet to ever-larger audiences was a kind of natural history of the nonliving world.[12] To capture what it meant to do natural history, in practice rather than in the abstract, we need to notice the measurements and experimental interventions of all types, along with the close description. For the Réaumur-inspired naturalist, the appeal was in the doing, in seeing the previously unseen, and in devising inventive and reliable ways of doing so. We should also notice that the big questions—generation, first and foremost, but also animal instinct, animal souls, materialism—were never far from the practice of naturalists, even if they did maintain a wary stance toward hypotheses and speculative theories. For many who read Réaumur's books and wrote to him with their own observations, pondering such questions was part and parcel of their fascination with the natural world and their own interventions in it.

The ancient distinction between natural philosophy, which purportedly searches for causes, and natural history, which describes and classifies, was fraught with tension by the eighteenth century and is perhaps not very useful as we parse scientific practice in this period. What was to count as a cause, after all? Réaumur refrained from theorizing, generally, but he was always interested in finding what he called causes: the mechanism of the caterpillar's jaws or the carnivorous bird's digestion, or the effect of humidity on embryonic development in the chicken egg. Physics, on this view, means the search for mechanisms, but these must be observable. For Buffon, physics meant something more like general, even abstract, laws, as Hoquet has pointed out.[13] Nollet's claim that the methods of physics were essential to the practice of natural history leads us to revise the notion that eighteenth-century life sciences developed chronologically from (mere) empiricism to theory, or from description of particulars to general principles. The physics-inspired natural history of Réaumur and so many others sought a kind of understanding, through repeated series of observations and experiments, that became part of daily life for many naturalists.

Following natural history as it developed in the generations after Réaumur would take us well beyond the scope of this book. It would be impossible to do justice to all the strands of observational and field sciences that unfolded in the latter part of the eighteenth century, some of them eventually feeding into the new discipline of biology in the nineteenth. Certainly those who studied insects and microorganisms continued to hold Réaumur up as an ideal observer; Senebier, Saussure, and Huber in Switzerland and Spallanzani in Italy are only a few examples of naturalists of later generations who continued along the lines we have been tracing.[14] With the increasing subdivision of life sciences into specializations around the turn of the nineteenth century, the broad and eclectic style of natural history might look very old-fashioned. But look again. Even as they specialized, field observers like the entomologists Henri Fabre and Jules Michelet were arguably following in the path of the kind of natural history that inspired Bazin, Lyonet, and the young Bonnet. Their lineage continues right down to the twentieth century with formicologists like William Morton Wheeler, who published Réaumur's work on ants from the manuscripts left in the academy in Paris, and who looked back to the eighteenth-century naturalist as his forebear. Wheeler, in the 1920s, was just as interested as Bazin, two hundred years earlier, in catching nature in the act.[15]

In our day, scientists who observe and experiment on animals, whether insects or fishes or birds or spiders, have subdivided into many disciplinary specialties: neuroethology, chemical ecology, biophysics, sociobiology, and so on. They use electron microscopes, high-speed video cameras, spectrographs, and any number of chemical tools—not to mention the basic evolutionary framework that informs all such work. Still, when you read in the daily newspaper that a neuroethologist from Lund University has been studying the preening behavior of dung beetles in hot climates by, among many other things, fitting them with silicone boots to keep their feet cool, you cannot help think of the chicken in Réaumur's yard with its blue stocking or the frogs in Charenton in their little waterproof pants.[16] Such ingenious interventions are rife in the many subdisciplines of field biology, where scientists still do their best to catch these creatures in the act.[17]

ACKNOWLEDGMENTS

Research for this book was supported in part by the National Science Foundation under grant number 0749082; by a Dibner Fellowship from the Huntington Library; and by the Committee on Research of the UCLA Academic Senate. I would also like to express my gratitude to Lorraine Daston for inviting me to be part of the research group on the history of scientific observation at the Max Planck Institute for the History of Science in Berlin. The congenial colleagues in this group, and the time I spent at the MPI in 2008, were helpful to my thinking about this project at a crucial time in its development.

I had the benefit of the goodwill and expert assistance of many librarians and archivists as I did my research. I must mention first of all the staff of the Archives of the Academy of Sciences in Paris, a most congenial place to work. The archivist, Florence Greffe, was always welcoming and facilitated my work in every possible way. I would also like to thank Alice Lemaire and her colleagues in the manuscript collections of the Natural History Museum in Paris; the librarians at the MPI, who were endlessly accommodating; and the staff of all the other libraries and archives where I found manuscript material. Closer to home, thanks to Katherine Donahue and Russell Johnson of the UCLA Biomedical Library, who were very helpful with obtaining photographic reproductions for my illustrations. I had invaluable advice from Adam Politzer on a variety of technical matters; he helped patiently with all kinds of hardware and software problems, and especially with everything having to do with the digital photography that was essential to my research.

Several friends read all or part of the manuscript at various times. I'd like particularly to thank Jane Smith, my Huntington Gardens companion and a most perceptive reader; Pamela Smith for her stalwart support and helpful comments over the years; Mimi Kim for her insights and her challenges; and Anne Secord for understanding and appreciating what I was trying to do. Ted Porter has been a most supportive and helpful reader throughout. My work has also benefited from conversations and consultations with Kapil Raj, Jim Secord, Otto Sibum, Marc Ratcliff, Lucia Dacome, Marco Bresadola, and

Paola Bertucci. In Paris, heartfelt thanks to Kapil Raj, Karine Chemla, and Bruno Belhoste for their frequent and generous hospitality and for listening enthusiastically to tales of my archival discoveries as they unfolded. In Geneva, thanks to Marc Ratcliff for his hospitality and his generosity with the results of his own research. I had essential help in navigating the notarial records at the Archives nationales in Paris from my UCLA colleague Kate Norberg, who shared her insider knowledge with me and smoothed my way through the *Minutier central*. Other colleagues generously shared documents, photographs, unpublished work, and essential references: François Pugnière, in Nîmes; Paola Bertucci; Virginia Dawson; Marie-Noëlle Bourguet; Andrew Curran; and Jean-François Gauvin.

I dedicate this book to my sons, Noah and Adam Politzer.

NOTES

Abbreviations

AN: Archives nationales

AN-MC: Archives nationales, Minutier central

Art de faire: R. A. F. de Réaumur, *L'art de faire éclorre et d'éléver en toute saison des oiseaux domestiques de toutes espèces*, 2 vols. (Paris: Imprimerie royale, 1749)

AAS: Archives, Académie des sciences, Paris

AAS-FR: Fonds Réaumur, Académie des sciences, Paris

AAS p-v: Procès-verbaux de l'Académie des sciences, Paris

BCMHN: Bibliothèque centrale du Muséum d'histoire naturelle, Paris

BG: Bibliothèque de Genève

Corr. inédite: Maurice Trembley, ed., *Correspondance inédite entre Réaumur et Trembley* (Geneva: Georg & Compagnie, 1943)

HAS: *Histoire de l'Académie royale des sciences*

HL: Houghton Library, Harvard University

Lettres inédites: G. Musset, ed., *Lettres inédites de Réaumur* (La Rochelle: Mareschal & Martin, 1886)

MAS: *Mémoires de l'Académie royale des sciences*

MI: R. A. F. de Réaumur, *Mémoires pour servir à l'histoire des insectes*, 6 vols. (Paris: Imprimerie royale, 1734–1742)

R: Réaumur

Chapter One

1 Lyonet, note to Lesser, *Théologie des insectes*, 1:128–29.

2 Spary, *Utopia's Garden*, 5–6, discusses the eclectic nature of natural history for the later period. See also Daston, "Attention and the Values of Nature."

3 On the consolidation of botany as descriptive practice in an earlier era, see Ogilvie, *The Science of Describing*.

4 On observation as epistemic genre, see Daston, "Empire of Observation," in Daston and Lunbeck, *Histories of Scientific Observation*.

5 Roger, *Sciences de la vie*; Roger, *Buffon*, chap. 5. On the meaning of "history" and "description," see also Stalnaker, *Unfinished Enlightenment*, chap. 1.

6 Roger, *Sciences de la vie*, 392.

7 Roger, *Buffon*, 107. Roger divides pre-Buffon naturalists into "*observateurs*"—and especially observers of insects—and "*classificateurs*," who are primarily botanists.

8 More recent scholarship has gone beyond Roger's idealist historiography. E. C. Spary has examined the institutional settings and patronage relations underpinning natural history operations at the Jardin du roi (later Jardin des plantes) in Paris, laying bare the dynamics of acquisition and cultivation that built a discipline, an institution, and individual careers. Spary, *Utopia's Garden*. Thierry Hoquet, in his reading of Buffon's work, shows areas of overlap in the perspectives of Réaumur and Buffon, especially in their shared objections to simplistic natural theology, breaking down some of the overly stark distinctions drawn by Roger. Hoquet, *Buffon*, esp. 525–33. The Aristotelian distinction between description (natural history) and the search for causes (natural philosophy) was already breaking down in the Renaissance. See Findlen, *Possessing Nature*.

9 Ratcliff, *Quest for the Invisible*. Ratcliff challenges the standard story about microscopy, according to which the eighteenth century was a period of stagnation between the efflorescence of microscopic discoveries in the seventeenth century and the technical advances of the nineteenth.

10 *MI*, 1:49–50.

11 See Pomata, "*Praxis historialis*," for the earlier use of the genre "historia" in medicine.

12 The canonical work of this type in French was the Abbé Noël Pluche's *Spectacle de la nature*, a book that went through countless editions, to the great profit of its author and his publisher. On Pluche, see Trinkle, "Pluche's *Le spectacle de la nature*"; Loveland, *Rhetoric and Natural History*, 73–74.

13 Lesser, *Insectotheologie* (Frankfurt, 1738). Other natural theological works by Lesser include *Lithotheologie* (Hamburg, 1735) and *Testaceo-Theologie* (Leipzig, 1744). On this prolific tradition of German physicotheology, see Clark, "Death of Metaphysics"; on English and French natural theology, see Roger, "Sciences de la vie," 242–49.

14 On Lyonet, see van Seters, *Pierre Lyonet*; Hublard, *Naturaliste hollandais*.

15 "C'est un ouvrage composé à l'Allemande, où le travail et la lecture ont plus de part que le génie et l'expérience." (Note that "*expérience*" could also be translated as "experiment.") Lyonet to R, 11 April 1743, HL, Ms. Fr. 99.

16 Lyonet, note to Lesser, *Théologie des insectes*, 2:118–19.

17 Ibid., 2:119.

18 Réaumur, "Sur les diverses reproductions qui se font dans les écrevisses, les omars, les crabes, etc.," *MAS* (1712), 223–43.

19 Lyonet, note to Lesser, *Théologie des insectes*, 1:125. Stigmata is an archaic term for the openings to the insect respiratory system; the modern term is spiracle.

20 Ibid., 1:148.

21 For an analysis of the rhetorical use of final causes in natural historical writing, see Loveland, *Rhetoric and Natural History*, chap. 2.

22 Daston, "Attention and the Values of Nature," 101.

23 On overlap in personnel, see Sturdy, *Science and Social Status*, 399–412. On tension between the institutions, see Guerrini, "Duverney's Skeletons." Spary points to Buffon's attempts to "lay claim to the sorts of natural knowledge that were being produced at the Academy of Sciences" in the 1740s. Spary, *Utopia's Garden*, 18–19.

24 See Bourguet, "Measurable Difference," for Réaumur's network of correspondents measuring temperature and collecting specimens in the colonies. On connections between the colonies and Paris more generally, see McClellan and Regourd, *Colonial Machine*.

25 This was not a concerted research program, as the unfinished project on animals had been in Perrault's day. Guerrini, "Duverney's Skeletons."

26 Founded in 1706 as a coequal of the Paris Academy of Sciences, the Société royale des sciences (Montpellier) thus had a rather different relation to the capital than other provincial academies. See Roche, *Siècle des lumières*. On its early history, Williams, *Cultural History of Medical Vitalism*, 26–31.

27 "Bon de Saint-Hilaire, François-Xavier," *Encyclopédie méthodique (Histoire)* (Paris: Panckoucke, 1784), 1:654. Bon had participated in writing the statutes of the Montpellier Société royale des sciences; some years later he became a member of the Paris Academy of Inscriptions.

28 Luigi Fernando Marsigli, *Histoire physique de la mer* (Amsterdam: Aux dépens de la compagnie, 1725). The experiments in Marseilles were done in 1706. On Bon's involvement with the controversy about the nature of coral, see McConnell, "Flowers of Coral." Bon's other scientific activities included keeping detailed meteorological records at his private astronomical observatory; he had a well-known collection of instruments, antiquities, and naturalia as well.

29 François-Xavier Bon de Saint-Hilaire, "Dissertation sur l'utilité de la soye des araignées," *Assemblée publique de la société royale des sciences* (Montpellier, 1710; reprinted in Paris, 1710).

30 François-Xavier Bon de Saint-Hilaire, "A Discourse upon the Usefulness of the Silk of Spiders," *Philosophical Transactions* 27 (1710–1712), 2–16 (http://www.jstor.org/stable/103101). The paper also appeared in Italian translation in the same year.

31 AAS p-v, 14 December 1709. Fontenelle mentioned Bon's status: "premier Président en la Chambre des comptes, aydes et finances de Montpellier." According to Réaumur, the Paris Academy regarded the mittens "with the pleasure provoked by curiosities [*les choses curieuses*]." R, "Examen de la soie des araignées," *MAS* (1710), 386.

32 Réaumur was appointed *élève* in the mathematics class in 1708; by 1709 he had begun his move into natural history with a paper on the growth of shells (*MAS* 1709).

33 Presented 12 November 1710 (AAS p-v).

34 R, "Examen de la soie des araignées," *MAS* (1710), 387.

35 Ibid., 390.
36 Ibid., 401–2.
37 Ibid., 405.
38 Ibid, 407.
39 Bon de Saint Hilaire to [Pitot], n.d., AAS-FR, Correspondance. The document seems to have been copied for Réaumur from a longer letter. Bon was an official correspondent of Henri Pitot, who was probably the recipient of the original letter.
40 Parennin reported to the Paris Academy of Sciences, after the death of the Chinese emperor, that he had translated Réaumur's paper into Manchu. See *HAS* (1726), for a report on Parennin's translation. Bon reprinted his original paper in 1726 in Avignon, and again in 1748 with a Latin translation; the later edition occasioned another lengthy review by the Jesuits.
41 Bon, "Utilité de la soie des araignées." Bon had used chemical analysis in his work on coral and applied the same distillation techniques to spider silk.
42 On the Baconian program of natural historical "matters of fact" as the basis of true knowledge of nature, see Pomata, "*Praxis historialis*," 113.
43 After the premature death of his brother in 1719, Réaumur supervised estate matters, spending the annual two-month academic vacation in Poitou.
44 On Bon's collections, see Dézallier d'Argenville, *Histoire naturelle éclaircie*, 211–12.
45 The survey continued for three years, from 1716 to 1718. The documents produced by this enterprise have been published in full in Demeulenaere-Douyère and Sturdy, *Enquête du regent.* On the thermometer, see Gauvin, "Instrument That Never Was." The *Description des arts et métiers* only saw the light of day after Réaumur's death, when it was taken over by Duhamel du Monceau; see Gillispie, *Science and Polity*, 337–56; Sheridan, "Recording Technology."
46 The only treatment of Réaumur's whole career and oeuvre remains Torlais, *Réaumur*; see also the collection Grassé, *Vie et l'oeuvre*. On his experiments with frogs, see Terrall, "Frogs on the Mantelpiece."
47 *MI*, 2:xxxi.
48 Ibid.
49 Harkness, *Jewel House*; Smith, *Body of the Artisan*.

Chapter Two

1 Charles Bonnet to R, 4 July 1738, AAS, dossier Bonnet.
2 Bazin to R, 4 August 1738, AAS, dossier Bazin.
3 *MI*, 1:538–39.
4 Ibid., 1:539–40.
5 Ibid, 2:101–2.
6 Ibid., 1:549.

7 Ibid., 1:608.
8 Ibid., 1:45.
9 See Terrall, "Following Insects Around."
10 Among Réaumur's assistants were Henri Pitot, Jean-Antoine Nollet, François-David Hérissant, Mathurin Jacques Brisson, François Menon, and one Dr. Baron. Pitot, Nollet, Hérissant, and Brisson all became regular academicians in due course; Menon was an official correspondent of the academy from 1747 until his death less than two years later.
11 The estates were occupied and managed by Réaumur's younger brother until his premature death in 1719.
12 Réaumur mentions that Baron lived with him: "M Baron, qui avant de s'établir Medecin à Luçon, avoit demeuré chés moi à Paris, & qui y avoit même eu soin de mes menageries d'insectes . . ." *MI*, 1:51.
13 Girard de Villars remarked on having seen Baron, who "is continuing to work, with gusto." Girard de Villars to R, 18 May 1733, BCMHN, Ms. 2624.
14 Baron to R, 2 May 1733, BCMHN, Ms. 2624.
15 Baron to R, 28 June 1733, BCMHN, Ms. 2624.
16 Baron to R, 2 May1733, BCMHN, Ms. 2624.
17 Baron to R, 2 May 1734, BCMHN, Ms. 2624.
18 Girard de Villars to R, 28 December 1732, BCMHN, Ms. 2624.
19 Ibid.
20 Girard de Villars to R, 18 May 1733, BCMHN, Ms. 2624.
21 Ibid. Villars referred to these insects as *blattes*.
22 Girard de Villars to R, 21 June 1734, BCMHN, Ms. 2624.
23 Many standard reference works say that Bazin was a doctor, but this is incorrect. The Strasbourg parish record of Bazin's death gives his professional status as *avocat au parlement*; copy in AAS, dossier Bazin. His widow's testament also gives this title; "Testament de Marie Mettra, veuve Bazin," 13 February 1769, AN-MC, ET/XLV/536. On the Paris legal profession, see Bell, *Lawyers and Citizens*. Bell points out (31) that many barristers never practiced law as such.
24 Bluche, "Officiers du grenier à sel," 298. Réaumur mentions Bazin's position at the salt depot; *MI*, 1:50.
25 Bazin's letters to Réaumur mention people in the household in Charenton, which he seems to have known well. Bazin and Réaumur lived in the same neighborhood from 1717 until sometime in the 1720s (addresses in *Almanach royal*). See also Terrall, "Following Insects Around."
26 Bazin to R, 13 May 1733, AAS, dossier Bazin: "Je me suis refugié." Bazin and his wife, Marie Mettra, did not seek a formal *séparation des biens*, so there was no court case. A dispute about property, possibly an inheritance, seems to have instigated the separation. Bazin died in 1754 in Strasbourg, Mettra in 1769 in Paris; her testament, cited above (AN-MC, ET/XLV/536), identified her as Bazin's widow.

27 *MI*, 1:51. Réaumur, who met with Bazin's wife at least once in Paris, maintained complete discretion about his friend's reasons for leaving the city.

28 Bazin to R, 19 March 1733, AAS, dossier Bazin: "L'honneur de vous ecrire sans vous parler d'insectes peu ou prou, ce seroit ecrire à sa maitresse sans luy parler de son amour." On the affective dimension of natural history observation, see Daston, "Attention and the Values of Nature."

29 "Une des choses des plus agreables est de trouver le giste des oeufs d'une infinité d'insectes. C'est si me semble un grand plaisir de découvrir toutes ces petites cachettes ou de surprendre les meres cachants leurs tresor. C'est bien la *prendre la nature sur le fait* [my emphasis]: car c'est une chose admirable de voir la varieté infinie des moyens que la nature a inspiré a ces animaux pour tranquiliser leur inquietude et leurs amour pour leurs petits, c'est aussy ce que je cherche avec le plus grand soin et dont j'ai bien de la peine a tenir a bout. Par exemple ce petit ver de fresne que j'ai trouvé sur la scrophulaire sortait de son oeuf, et cet oeuf estoit attaché au plus jeune point de la plante et avoit l'air d'y avoir esté implanté tout nouvellement. Il n'y avoit nulle apparence qu'il fut sorti avec le rejetton comme Swammerdam dit que cela arrive quelque fois. J'aurois grand besoin de vos yeux Monsieur dans ces cas la." Bazin to R, 29 June 1733, AAS, dossier Bazin.

30 Though Bazin did have some private income from annuities, Réaumur loaned him money in this period, and consulted with his Paris lawyer and gave advice about Bazin's legal and financial situation. See Bazin to R, 8 June 1733, AAS, dossier Bazin.

31 Bazin to R, 13 May1733, AAS, dossier Bazin.

32 Bazin to R, 8 June 1733, AAS, dossier Bazin.

33 Bazin to R, 29 June 1733, AAS, dossier Bazin.

34 Bazin to R, n.d. [early December 1734?], AAS, dossier Bazin.

35 Bazin to R, 11 December 1734, AAS, dossier Bazin.

36 Réaumur covered the many species of inchworms (*arpenteuses*) in chapters 8 and 9, *MI* 2:323–90.

37 Bazin (from Strasbourg) to R, 12 January 1735, AAS, dossier Bazin.

38 Bazin to R, 15 August 1736, AAS, dossier Bazin. I have translated *chenille du gazon* as "lawn caterpillar."

39 Bazin to R, 8 June 1733, AAS, dossier Bazin.

40 Ibid.

41 Vivant was *évêque suffragant* in Strasbourg; he also held a benefice as bishop of Paros (or Pharos), so Bazin called him "M. de Paros." In the aftermath of his separation from Marie Mettra, Bazin told Réaumur of a "secret" offer from Vivant to come to Strasbourg. Bazin to R, 19 June 1733, AAS, dossier Bazin. It took another year to settle his legal and financial affairs, and he accepted Vivant's offer in spring 1734. Bazin to R, May 1734, AAS, dossier Bazin.

42 After Vivant's death in 1739, the Cardinal de Rohan (bishop of Strasbourg) ar-

ranged for Bazin to catalog Vivant's extensive library, and then gave him a position as his own librarian. Bazin lived in Rohan's establishment (in the new episcopal palace in Strasbourg) until his death in 1754.

43 Swammerdam had pioneered these anatomical techniques on insects in the seventeenth century. See Swammerdam, *Histoire générale des insectes*. Bazin's study of anatomy in Strasbourg may be the source of the erroneous claim in biographical dictionaries that he practiced as a doctor, or that he had a degree from the medical faculty in Strasbourg.

44 Réaumur did only simple dissections himself. He sent Bazin the "secret" recipe, used by the Paris anatomist Jacques Winslow for injecting vessels in anatomical preparations, which Bazin adapted successfully to insects. Bazin to R, 22 March 1738, AAS, dossier Bazin.

45 Bazin to R, 9 December 1737, AAS, dossier Bazin.

46 "This *tablette* is not my apprenticeship. I practiced earlier on another that I managed less well." Ibid. Réaumur showed Bazin's caterpillar preparations to the Academy of Sciences, and then added them to his insect collection. *MI*, 4:251.

47 Bazin to R, 28 August 1737 (sent shortly before Réaumur's annual departure for Poitou): "Je me sens tout plein de bonnes choses a vous dire. Je vous conseille cependant d'en remetre la lecture pour quelque auberge ou vous vous trouveréz desoeuvré."

48 Bazin to R, 26 May 1735, AAS, dossier Bazin.

49 Ibid.

50 Ibid.

51 Bazin to R, 3 June 1735, AAS, dossier Bazin.

52 Bazin to R, 28 June 1737, AAS, dossier Bazin.

53 Bazin to R, 15 July 1738, AAS, dossier Bazin.

54 Bazin to R, 4 August 1738, AAS, dossier Bazin.

55 Bazin to R, 29 November 1737, AAS, dossier Bazin.

56 Ibid.

57 Raoul to R, 1 August 1732 (HL, Ms. Fr. 99), written from his country house in St. Aubin en Blanquefort. He refers to the freeze of 1709, when the frost killed two thousand pine trees. Raoul to R, 11 December 1734, AAS-FR, Correspondance. I have not been able to find biographical information on Raoul, including his full name.

58 Raoul to R, 10 June 1732, HL, Ms. Fr. 99.

59 Ibid. Raoul thanks Réaumur for the remedy for clothes moths, recently published in the *Mémoires* of the Paris Academy: "Grace à l'huile de therebentine que vous nous avés indiqués je ne vois plus mes habits devorés par les teignes." See R, "Histoire des teignes" (two-part memoir).

60 Raoul recalled an argument he had had in 1692 in Paris with a nobleman about whether beehives were ruled by kings. Raoul to R, 2 October 1740, AAS-FR, Correspondance.

61 I translate Raoul's *poux des bleds* as "grain weevil." Raoul's first contact with Réaumur predates by three years the publication of the first volume of the latter's insect book.

62 Raoul to R, 10 June 1732, HL, Ms. Fr. 99. He mentioned that he had long been familiar with Réaumur's work and had repeated his observations many times. Two letters signed "Pradat, curé de Begadan" survive in Houghton Library: Pradat [Raoul] to R, 1 January 1731 and 1 February 1732, HL, Ms. Fr 99.

63 Pradat [Raoul] to R, 1 January 1731, HL, Ms. Fr 99.

64 Raoul to R, 10 June 1732. White vitriol (*vitriol blanc*) is zinc sulfate in modern terminology.

65 Pradat [Raoul] to R, 1 January 1731, HL, Ms. Fr 99.

66 Raoul to R, 23 May 1733, BCMHN, Ms. 2624.

67 Ibid.

68 Raoul to R, 16 January 1734, HL, Ms. Fr. 99.

69 Raoul to R, 3 April 1734, BCMHN, Ms. 2624.

70 *MI*, 2:149. Bonnet also worked on pine caterpillars in 1738, probably inspired by Réaumur's account in *MI*, vol. 2 (1736). Cramer to Jallabert, 24 December 1738, BG, Ms. SH 242.

71 Ibid., where Réaumur explains that he had first learned about pine-caterpillar silk from a 1710 publication by "two medical students in Montpellier."

72 Ibid., 2:151. Raoul mentioned the idea of making stockings in his letter of 16 January 1734 (HL, Ms. Fr. 99).

73 See *MI*, 1:150 ff.

74 Ibid., 2:157.

75 Raoul to R, 17 April 1734, BCMHN, Ms. 2624.

76 For other examples, see Terrall, "Following Insects Around."

Chapter Three

1 See R (from Charenton) to Crousaz, 2 June 1718, Bibliothèque universitaire de Lausanne, Fonds Crousaz, box IX: "Pour jouir de la belle saison, et pour travailler avec quelque tranquillité, j'ay presque toujours resté à la campagne depuis les festes de Pasques. Je ne me rends à Paris que les jours d'Academie." In the 1730s, Réaumur's movements between Paris and Charenton can be followed in his record of thrice-daily thermometric measurements. AAS-FR, dossiers 48–50.

2 Maistre, *Les précieuses*. See also Harth, *Cartesian Women*. Réaumur describes the house as "si celebré dans les lettres de Voiture"; R to Caumont, 23 July 1736, BG, Ms. Trembley 5. The Hôtel d'Uzès was on rue Saint-Thomas-du-Louvre, which ran between the Seine and the rue Saint-Honoré. The street was destroyed in the 1850s to make way for an expansion of the Louvre.

3 For a wealth of detail about the architecture and ownership of the house, especially in its glory days in the seventeenth century, see Babelon, "Hôtel de Rambouillet." Babelon reconstructs the building from archival documents; his architectural rendering is on p. 327. Babelon did not realize that Réaumur had occupied the premises for eight years.

4 "Ouvrages de fer et d'acier," *Mercure de France*, December 1726, 2729–33 (quotation on 2729). (There was also a short announcement in the June 1726 issue.)

5 R, "Idée générale du nouvel art d'adoucir le fer fondu," AAS p-v, 4 May 1726. The passage continues: "Je travaillois à le faire mettre en pratique, et je cherchois à le perfectionner. Il n'est que ceux qui ont déja mis la main à l'oeuvre qui sachent combien les Experiences necessaires se multiplient dès qu'on veut approfondir une matiere qui a semblé en demander quelques-unes." This paper was not published.

6 R, *Art de convertir*; summary in *HAS* (1722), 39–55. Réaumur presented a series of papers on cast iron to the academy between 1722 and 1726; most were only published in the posthumous second edition of his book, with the new title *Nouvel art d'adoucir le fer fondu* (Paris: Imprimerie royale, 1763). For the contents and publication history, see Smith, *Réaumur's Memoirs on Steel*, appendix B.

7 The document (dated 22 December 1721) establishing the pension, payable each year from the *ferme des postes*, is reproduced in "Brevet de pension de M. de Réaumur," *Archives historiques de Saintonge et de l'Aunis* 15 (1887): 23–24.

8 R (from Cosne) to Mazzoleni, 27 May 1724, Biblioteca Laurenziana (Florence), Ms. Ashburnham 1522. He was there again for five weeks the following year: R to Crousaz, 22 May 1725, Bibliothèque universitaire de Lausanne, Fonds Crousaz, box II.

9 Pierre Jarosson (from Cosne) to R, 19 November 1723, AAS-FR, Correspondance. On the history of this enterprise, see Bontemps and Prade, "Magasin parisien." On the original privilege, the primary investor was Chevalier de Bethune; he had eleven associates, including Jarosson.

10 For detailed information about the operation, see "Visite des fourneaux et forges à l'hotel d'Uzès, 1 July 1726," AN, Z/1j/581. The owner of the house, worried about the danger of fire, requested the expert evaluation of the situation. The enterprise employed forty workers at the Hôtel d'Uzès. Bontemps and Prade, "Magasin parisien," 223–26. A few of the ornamental door knockers can still be seen in their original locations in Paris.

11 Ibid.

12 See also Pitot to R, 29 December1743, AAS, dossier Pitot. Réaumur was the godfather of Pitot's son René.

13 Before moving to the Louvre in 1699, the academy's premises in the Bibliothèque du roi included a laboratory. During the Enquête du regent (1716–1718), Réaumur supervised the assaying of mineral specimens in the "laboratory of the academy," though its location is not specified in the documents. For possible locations, including the hypothesis that it was identical to Réaumur's laboratory, see Demeulenaere-Douyère and Sturdy, *L'enquête du regent*, 38. Academician-chemists, some of whom were pharmacists as well, generally had laboratories in their homes or in the homes of their patrons.

14 Pitot entered the academy in 1724 as *adjoint mécanicien.* For Pitot's biography and early career, see Grandjean de Fouchy, "Eloge de M. Pitot," *HAS* (1771), 143–57: "The academy had entrusted the direction of its laboratory to M. de Réaumur, and he always used the small amount of revenue attached to it to employ some young man whose talents he knew and whom he was grooming [*formoit*] for the academy" (146). According to Fouchy, Pitot also worked on various parts of the *Description des arts et métiers*, none of which was published in Réaumur's lifetime. Pitot's own work was initially in mathematics and then in mechanics, especially hydraulics. Eventually he ended up as the director of the Canal de Languedoc.

15 Pitot continued to work with Réaumur off and on in the early 1730s, while advancing his own career in the academy through his research on mathematics and engineering.

16 Nollet called the adoption of the Réaumur thermometer a "*révolution presque totale*"; the instrument displaced most competitors in a short time, at least in the French market. Nollet, *Leçons de physique expérimentale*, 4:397.

17 *Explication des principes établis par M. de Réaumur, pour la construction des thermomètres dont les degrés soient comparables* n.p., n.d. [ca. 1730], quotation on 2–3. This pamphlet was written for nonspecialists wishing to purchase and use the new thermometers; corrected manuscript in Réaumur's hand in AAS-FR, dossier 47.

18 On the ambiguity of "boiling water" in this context, see Gauvin, "Instrument That Never Was."

19 A short version of this paper was read at the public session of the academy in November 1730; the full text stretched over four sessions in January 1731, with the second paper read in June. R, "Règles pour construire . . ." *MAS* (1730), 452–507; R, "Second mémoire sur la construction des thermomètres, dont les degrès sont comparables, avec des expériences et des remarques sur quelques propriétés de l'air," *MAS* (1731), 250–96. On the history of thermometry, Middleton, *History of the Thermometer*; Birembaut, "Contribution de Réaumur."

20 R, "Essais sur le volume qui résulte de ceux de deux liqueurs mêlées ensemble," *MAS* (1733), 167–68.

21 Messier, "Mémoire sur le froid extraordinaire que l'on ressentit à Paris . . . au commencement de cette année 1776," *MAS* (1776), 1–155 (transcription of

label on a thermometer made by Pitot in 1730 for Gravet de Livry on 140). See also Birembaut, "Contribution de Réaumur." Pitot served as one of the official examiners of the instrument when it was submitted to the academy for examination; not surprisingly, the report was favorable. On this early history of Réaumur's thermometer, see Gauvin, "Instrument That Never Was," Pitot on 22n80. Pitot and Nicole's report on the thermometer is in AAS p-v, 1731, f. 110.

22 Weather diaries recording temperatures taken two or three times daily survive in the archives from 1732 through 1747; some of this data was published in *MAS* starting in 1733. On the meanings and uses of weather diaries, see Golinski, *British Weather*.

23 See Bourguet, "Measureable Difference," for the use of Réaumur thermometers in the next generation.

24 Baron (from Luçon) to R, 2 May 1733, BCMHN, Ms. 2624.

25 Grandjean de Fouchy, "Eloge de M. l'Abbé Nollet," *MAS* (1770), 122.

26 Nollet made a pair of globes, dedicated to the Duchesse du Maine, in 1728. Ronfort, "Science and Luxury." See also Pyenson and Gauvin, *Art of Teaching Physics*.

27 Nollet kept in his possession a thermometer made with his patron in 1732, and claimed it was still accurate almost forty years later. He left this instrument in his will to the academician Brisson, assistant to Réaumur from 1749 to 1757. For the construction date of this thermometer, see Nollet, *Art des experiences*, 3:182. See also Gauvin, " Instrument That Never Was."

28 Nollet, *Art des expériences*, 3:144. Réaumur's first thermometric scale had started from freezing point of water, which differs slightly from the melting point of ice and is more difficult to ascertain unequivocally.

29 Nollet, *Programme*, thermometers on 185–87.

30 Jean-François Séguier, "Récits de voyage." Bibliothèque municipale de Nîmes, ms. 129. I thank François Pugnière for generously making this document available to me. Séguier remarked that most of Réaumur's experiments were done in the laboratory at Charenton. Séguier moved in the same social circles as Réaumur, and later exchanged specimens, letters, and books with him (see chapter 6).

31 Ibid.: "J'y vis une très grande quantité d'insectes et de papillons conservés entre deux verres qui étaient collés par les extrémités et les animaux étaient préparés de façon à ne pas craindre les mites."

32 *MI*, 2:403. Nollet's work as insect-keeper is also mentioned on *MI*, 3:178.

33 "Experiences à faire faire par M. l'Abbé Nollet," AAS-FR, carton 5, dossier 47, f. 46. This list can be approximately dated by comparison to other manuscript notes on experiments with weevils (*charençons*) dated throughout 1733 and 1734. The experiments on artificial ice were presented to the academy in 1734. R, "Congelations," *MAS* (1734).

34 "The weevils that I had left in Charenton in the laboratory, which is extremely humid, have multiplied there more than anywhere else, it seems to me. . . . From which it seems that the humidity that destroys the grain is very favorable to them." "Histoire des charençons," ms. notes in R's hand, March 1733, HL, Ms. Fr. 99.

35 *MI*, 6:131 (plate XII, figs. 9 and 11). I have not been able to trace Perreau's biographical details. He may have lived at Charenton: at one point Bazin asks Réaumur to "have M. Perreau visit the nut trees at Charenton" to find specimens of a certain caterpillar. Bazin to R, 13 May 1733, AAS, dossier Bazin. Perreau later worked in the south; he sent insects for R's collection. Guettard, "Inventaire du cabinet de M. de Réaumur," BCMHN, Ms. 1929 (iii), f. 135v.

36 R, "Maniere de faire perir les charençons, et peut-etre touts les autres insectes, qui nous incommodent," autograph notes, HL, Ms. Fr. 99. A reading of 40 degrees on Réaumur's instrument would correspond approximately to 50 degrees Celsius.

37 R, "Pour faire perir les insectes domestiques, par le chaud," 13 May 1736, autograph notes, HL, Ms. Fr. 99. He had worked on the clothes moths a few years earlier: R, "Histoire des teignes" and "Suite de l'histoire des teignes."

38 *MI*, 2:8.

39 Ibid., 2:12.

40 Ibid., 2:16–17.

41 Ibid., 2:13. The glass balls were probably supplied by the same artisans who made the glassware for his thermometers.

42 Ibid., 2:14.

43 Ibid., 2:11. "*Curieux en insectes*" could also be translated as "insect connoisseur."

44 Bazin to R, 3 June 1735, AAS, dossier Bazin. Bazin says he was "ravi en extase." Réaumur mentions Bazin's trials in *MI*, 2:12.

45 For an example of Réaumur's clumsy drawing, and a discussion of this point, see Daston and Galison, *Objectivity*, 87.

46 "Pour suppléer à ce qui me manquoit, j'avois fait instruire un jeune homme . . ."; *MI*, 1:54.

47 None of the documents mention his first name. The Regnaudins of La Rochelle were a branch of the family of Réaumur's father. The artist, and his residence in the house, is mentioned in *MI*, 1:54. Regnaudin was not named in print but can be identified from the invoice for his drawings: "Mémoire des desseins faits par le Sr. Regnaudin pour entrer dans les mémoires de l'Academie, et pour servir à une histoire des insectes entreprise par Mr. de Réaumur," AAS, Fonds Lavoisier, comptabilité, 1065 ac 1731. Regnaudin was paid less than the going rate per sheet of drawings. All of his drawings were of insects except for one of a bird's nest sent from India, and one sheet of thermometers. Some of the insect

drawings were engraved for papers in the *Mémoires* and for the first volume of the insect book.

48 In 1733 (after the death of Regnaudin), Simonneau was paid a total of 1,320 *livres* by the academy for sixty-five sheets of drawings. "Mémoires des desseins faits par le Sr. Simonneau . . ."; AS, Fonds Lavoisier, 1066 ac, 1068 ac, 1069 ac.

49 *MI*, 2:493.

50 The exact typographic representation of the elliptical reference varies slightly from volume to volume. The plates made from these drawings have only the name of the engraver, with no attribution for the drawing. When engraving from his own originals, Simonneau's signature indicated that he had made both drawing and engraving: "del et sculp."

51 *MI*, 1:55. Invoices for Simonneau's drawings, made "sous les yeux de M. de Réaumur," in AAS, Fonds Lavoisier, 1066 ac, 1068 ac, 1069 ac.

52 Madeleine Basseporte, the artist who illustrated Pluche's *Spectacle de la nature*, is an interesting comparative example. Trained by Claude Aubriet, "painter to the king," at the Jardin du roi, she signed her work and earned her living from her drawings. See Daston and Galison, *Objectivity*, 89.

53 Unpublished drawings in AAS-FR and in BCMHN, Mss. 1901 and 972.

54 Hélène Dumoustier was named *légataire universelle;* Pierre Jarosson was designated executor of the will (though he predeceased Réaumur by two years). Réaumur had no siblings, children, nieces, or nephews. A codicil dated a few years later mentioned a cousin, Mme. Nantia, to whom he left his portrait as a token. She was later the main challenger to the will, pursuing a long legal battle for the assets left in the will to Hélène Dumoustier.

55 "Testament de M. de Réaumur," in Caullery, *Papiers laissés par de Réaumur*, 7–11, quotation on 8. Réaumur did not deposit his testament with a notary, for some reason, leaving it open to challenge by various distant relatives.

56 Departing from eighteenth-century usage, I refer to Hélène Dumoustier by her last name in what follows, using the most common spelling of her family name (as it appears, e.g., in the record of her birth). In legal documents she was either Mlle. Hélène Dumoustier or Mlle. Dumoustier de Marsilly (to distinguish her from her maiden sisters); her mother, as a widow, was Hélène Potel (her maiden name), veuve Dumoustier. In correspondence, her name was spelled variously.

57 *MI*, 1:55–56.

58 "Testament de M. de Réaumur," in Caullery, *Papiers laissés par Réaumur*, 7–11.

59 For the spurious connection to Marsigli, see *Corr. inédite*, 327, ed. note. Hamonou-Mahieu repeats this and calls Hélène Dumoustier's mother "la comtesse de Marsigli." Hamonou-Mahieu, *Claude Aubriet*, 125. The equally spurious connection to Fontenelle is in Madeleine Pinault-Sorensen, "A propos de

relations et de correspondents de Cideville," *Revue Fontenelle* 5 (2008): 212. Torlais describes Dumoustier as "la collaboratrice, ou pour être plus moderne, la secrétaire de Réaumur": Torlais, *Réaumur*, 132. . I have found no evidence that she performed the duties of a secretary.

60 Depot de testament de Dlle. Dumoustier de Marsilly, 12 January 1766, AN-MC, ET/CXV/774.

61 In 1757, a document filed to protest her treatment during the official sealing of her residence (*apposition des scelles*) gave her age as forty-eight (she was actually sixty-six at this time). In 1759, when asked to state her name and age, she gave her age as "around sixty-four years"; she was then actually sixty-eight. An extract of her baptismal record was made when she invested in a tontine: AN-MC, ET/CXV/834 (19).

62 Both terms are very general. Many "bourgeois de Paris" lived from investments in municipal bonds and other financial instruments, though there is no evidence for this in Nicolas Dumoustier's case. At the time of Hélène's birth, the family lived in the rue de la Chanverrerie, Saint Eustache parish, where they still lived in 1701 (AN-MC, ET/XXX/159).

63 Réaumur mentions that she uses both names. R to Ludot, 2 July 1753, British Library, Add. 39757. Usually such a name would come from an estate or a title, but I found no mention or record of ennoblement or seigneurial property in her family.

64 "Création de pension," 22 July 1701, AN-MC, ET/XXX/159. The sisters were eighteen and twenty years old at this time. On choir nuns (*religieuses de choeur*), see Choudhury, *Convents and Nuns*, 20.

65 "Rente perpetuelle sur la ville de Paris," in notary's summary index, MC ET/CXXI/RE, 7 January 1721. The beneficiaries of the annuity are named in a later document. "Quittance 24 juillet 1761", AN-MC, ET/CXV/742.

66 "Interrogatoire d'Hélène Dumoustier de Marcilly," 22 November 1759, AN, Y 13951. All of the questions and answers quoted in the following pages, recorded in the third person by the notary, come from this previously unnoticed document.

67 Some of the details are garbled in the interrogation record, though Dumoustier says clearly in response to question 10 that Jarosson was the nephew of Hélène Potel, her mother. The family ties are documented more fully in Jarosson's marriage contract: "Mariage, Pierre Jarosson and Marie Magdeleine Fantel de Lagny," 20 March 1742, AN-MC, ET/CXV/533.

68 At Jarosson's marriage in 1742, Dumoustier, her sister, and her mother represented his side of the family as witnesses. The contract calls the sisters "*tantes à la mode de Bretagne*." AN-MC, ET/CXV/533. Nicolas Dumoustier's sister, Magdeleine, left a trace in the inventory of papers in Réaumur's study: "Magdeleine Dumoustier, veuve Thibaule." Thibaule was Jarosson's uncle, the brother of his

mother (Jeanne Elisabeth Thibaule). Hélène Dumoustier referred to Jarosson as her mother's nephew. "Interrogatoire," question 10, AN, Y 13951.

69 "Vos dames de Charenton": Pierre Jarosson to R, 19 November 1723, AAS-FR, Correspondance. Many of Réaumur's correspondents used the phrase "vos dames" to refer to the Dumoustier women; e.g., Granger to R, 5 August 1736, AAS-FR, Correspondance; Baron to R., 6 April 1744, AAS-FR, Correspondance.

70 "Etat civil de Réaumur," 6 October 1723. A photocopy of this document was kindly provided to me by Maryvonne Validire, curator of the Manoir des sciences, Réaumur. Other family names on the document: Brisson (Catherine Brisson was the widow of Réaumur's brother), Regnaudin.

71 In 1759 she recalled their move as taking place "about twenty-eight years ago." This would have been a sublease, as Réaumur was renting the whole property. "Bail, R. A. F. de Réaumur et Helen Potel, v[euv]e Nicolas Dumoutier," 4 April 1735. AN-MC, ET/CXXI/301. (Potel's name is variously spelled in the archives as Potel, Postel or Pautel.)

72 R to Caumont, 23 July 1736, BG, Fonds Trembley 5. R comments to Caumont on the move: "l'Hôtel Rambouillet, si celebré dans les lettres de Voiture . . . j'ai été obligé d'en sortir avant la fin de mon bail. Je me suis transporté depuis peu, avec touts mes cabinets dans la rue neuve Saint Paul. un demenagement de cette espece est une grande affaire." Réaumur's lease of a house in rue neuve Saint Paul, AN-MC, ET/ CXXI /304 (23 April 1736).

73 "Bail," 6 April 1740 (lease on rue de la Roquette house). AN-MC, ET/XCIV/220.

74 Bazin to R, January 1733, AAS, dossier Bazin.

75 "Mon epouse prie mademoiselle du Moutier de luy acheter une blonde toute montée a la mode du prix de 15#." Baron to R, 12 December 1734, AAS-FR, Correspondance.

76 Visits to Malnoue, five leagues from Paris, are documented in R's daily thermometric records, AAS-FR, dossiers 48–50. See also *MI* 4:507–8 for observations made on insects infesting cattle in the abbey's herd.

77 The widow Dumoustier died in 1743, in the house in rue de la Roquette: "Notoriété," 16 August 1743, AN-MC, ET/LXXVII/231.

78 After their mother died, Réaumur tried to improve the sisters' financial situation by negotiating a pension from the royal finance minister, who had apparently given reason to hope that he might assign them the income from a tax farm, though nothing came of this. R to [Moreau de Séchelles (finance minister, 1754–1756)], n.d., in *Lettres inédites*, 169–70: "Il m'est bien agréable de penser que j'obéis à vos ordres en vous indiquant des moyens de changer le sort de Mlles. du Moutier. L'intérêt que vous leur avez donné lieu d'espérer dans les sous-fermes, opérera sans doute ce changement, mais ce ne peut etre que dans

deux ans, et deux ans sont bien longs à passer quand on est dans la situation où elles se trouvent."

79 AN, Y 14677 (collection of documents related to inventory of Réaumur's property after his death). She was asked similar questions several times: "[Elle] dit que depuis qu'elle demeure en la maison ou nous sommes et ou est decedée la d[it]e. sa mere et une dlle. de ses soeurs, elle n'a point payé de loyers, qu'elle n'estoit point en pension chez led. deffunct, qu'elle tenoit son menage et qu'elle ne mangeoit chez M. de Réaumur que quand il y mangeoit et l'en prioit, et que les meubles qui sont dans les lieux qu'elle et la dlle. sa soeur occupoient en lad. maison leur appartiennent et offre pour cet effet de nous faire voir l'appartement et sa cuisine."

80 "Interrogatoire," 22 November 1759, "Pour Made. de Nantia contre la Dlle. de Marcilly," AN, Y 13951/B.

81 Ibid. "Les deniers du roy"; the Academy of Sciences reimbursed Réaumur for the cost of drawings and engravings for his works.

82 Réaumur reported the sister's death in 1749, to Trembley and to Bazin, both of whom knew her. Bazin sent letters to the Dumoustier women under cover of his letters to Réaumur, e.g., Bazin to R, 3 June 1735, AAS, dossier Bazin.

83 Detailed invoices for Dumoustier's drawings for the period 1740–1747, AAS-FR, dossier 61 (in R's hand). Copies of invoices from 1736–1738 in AAS, Fonds Lavoisier, 1080 ac, 1084 ac,1086 ac, 1090 ac, 1093 ac, 1095 ac.

84 "Testament de Réaumur"in Caullery, *Papiers laissés par Réaumur*, 9: "Je dois meme lui rendre ici une justice que sa modestie ne m'a pas permis de lui rendre dans mon ouvrage, autant que je le devais et que je l'eusse souhaité. C'est qu'elle m'a fourni plusieurs observations qui sont entrées dans le corps de mon ouvrage et *qui m'auraient échappé*. Touts ceux qui la connoissent sçavent d'ailleurs combien elle est estimable pour la douceur et la sagesse de ses moeurs et par les qualités de l'esprit et du coeur et qu'elle eut mérité d'etre mieux traitée par la fortune qu'elle ne l'a été."

85 *MI*, 3:447: "Cette galle fut d'abord observée par Mlle du ***." Note that the rendering of the veiled name has changed slightly.

86 Ibid., 3:511–12: "Dans une promenade dont étoit Mlle *** à qui je dois tant de beaux desseins, nous nous obstinâmes à chercher les insectes de l'intérieur de ces galles. J'en ouvris plusieurs sans y rien trouver, & Mlle *** découvrit un ver dans une des premieres qu'elle ouvrit; elle & moi nous en trouvâmes ensuite dans presque toutes les galles de cette espece que nous examinâmes; nous n'en avons vû qu'un seul en chacune, quoiqu'il y ait apparence qu'il y est en compagnie. . . . Nous en voyions souvent un, & nous étions incertains si c'étoit un ver, jusqu'à ce qu'il lui plut de se mettre en mouvement."

87 Ibid., 3:494.

88 "Interrogatoire" (see n80), question 21.

89 *MI*, 2:154.

90 Ibid., 2:155–56.
91 Ibid., 2:156.
92 For more on these observations, see Terrall, "Frogs on the Mantelpiece."
93 Swammerdam's observations of frogs appeared posthumously in 1737 in *Bybel der nature*, edited by Hermann Boerhaave. Before Boerhaave acquired Swammerdam's manuscripts, they were in the possession of Joseph Duverney, in Paris, and Réaumur evidently knew the observations and the images before they were printed. In observation notes dated 1736, he mentions Swammerdam's depictions and notes that others had also tried to reproduce the Dutch observations. "Grenouilles (1736)," autograph ms., AAS-FR, dossier 35.
94 Jean-Antoine Nollet to Lazzaro Spallanzani, 24 September 1768, in Spallanzani, *Carteggi*, 6:207–8. Nollet recalled that this was "about thirty years ago." Spallanzani reported his own experiments, including the artificial insemination, in *Expériences pour servir à l'histoire de la génération des animaux et des plantes*, trans. J. Senebier (Geneva: Chirol, 1785), 12.
95 On Spallanzani's experiments, see Terrall, "Frogs on the Mantelpiece," 199–202.
96 The large container held "twelve or thirteen" mating pairs; quotation from Réaumur, ms. notes, "Grenouilles, 1737." All other details are taken from notes dated 8 April 1740. Both documents in AAS-FR, dossier 35.
97 See Condorcet, "Eloge de Guettard," *HAS* (1787), 47–62. Guettard lived in Réaumur's household from 1740 to 1744.
98 "Grenouilles," 8 April 1740, AAS-FR, dossier 35.
99 *MI*, 5:545.
100 Ibid., 5:547.
101 Ibid., 5:504.
102 "Araignées: leur vol," (mss. notes), AAS-FR, dossier 29.
103 See testament of Lelarge de Lignac (1762), where he bequeathes her a small sum. AN, MC ET/CXV.
104 Bazin to R, n.d. (ca. Sept 1737), AAS, dossier Bazin.
105 Letters to Réaumur from people who visited, whether in Paris or Poitou or Charenton, show that the women were included in social gatherings as well as experiments: Trembley, Ludot, Bazin, and Lignac, among others, mention them frequently.
106 *MI*, 6:469.
107 Ibid., 6:481.
108 Ibid., 6:483.
109 Ibid., 6:484.
110 This was exactly the problem Réaumur encountered in his contemporaneous investigations of frogs—he could not see the male fertilize the eggs. See Terrall, "Frogs on the Mantelpiece."

111 *MI*, 6:489: "None of the insects known to me carry out such a huge operation, that must take so much effort . . . with such ease and quickness. The tub I spoke of, and others that I likewise kept filled with clods of earth inhabited by nymphs, made it possible to observe what I could not have seen in the river."

112 Ibid., 6:490.

113 For an overview, see Cooper, "Homes and Households." See also Schiebinger, *Mind Has No Sex?*; Hunter and Hutton, *Women, Science and Medicine*; Findlen, "Science as a Career in Enlightenment Italy"; Iliffe and Willmoth, "Astronomy and the Domestic Sphere"; Harkness, "Managing an Experimental Household." For an earlier period, Algazi, "Scholars in Households." On the location of science in the home, see also Shapin, "House of Experiment."

Chapter Four

1 *MI*, 1:51–52.

2 There was also an Amsterdam edition, in duodecimo format, with each volume printed in two parts (vols. 1–3, 1737; vol. 4, 1740; vol. 5, 1741; vol. 6, 1743). The projected seventh volume covered beetles and ants. R, *Natural History of Ants*; Caullery, *Papiers laissés par Réaumur*.

3 *MI*, vol. 1, "Avertissement," n.p.: "Je serai en état de profiter, pour les Volumes qui doivent suivre celui-ci, des lumieres qu'on voudra bien me communiquer. Je serai en état d'éclaircir, de rectifier, de corriger ce qui aura paru demander à l'être."

4 Fontenelle, report on *MI*, vol. 2, *HAS* (1736), 8–9.

5 *MI*, 3:xxxii.

6 Senebier, for example, in his book on the observational method in natural history, used Réaumur as his exemplary observer. Senebier, *Art d'observer*. Parts of the books (e.g., on bees) were translated into other languages. For information on translations, see the annotated bibliography of Réaumur's works by W. M. Wheeler, in R, *Natural History of Ants*, 263–74.

7 Réaumur discusses his stylistic choices in *MI*, 1:25. On styles of persuasion in the Academy of Sciences, see Licoppe, *Formation*.

8 *MI*, 1:25–26: "Plus les faits sont singuliers, plus ils demandent à être attestés. Celui qui les annonce pour la premiere fois, ne sçauroit trop assurer qu'il les a vûs, & comment il les a vûs: il n'y a gueres que dans des Mémoires où l'on puisse parler souvent sur ce ton. Quand on me rapporte que dans chaque ruche, dans chaque république d'Abeilles, il n'y a que quelques reines ou femelles, je ne suis pas assés persuadé si je soupçonne qu'on ne me parle que sur un oui-dire: je ne le serai pas même assés, si on se contente d'avancer qu'on l'a observé; je puis me défier de la maniere dont l'observation a été faite."

9 Fontenelle, *HAS* (1736), 8–9.

10 *MI*, 3:xxx.

11 Billate to R, 12 September 1739, AAS-FR, Correspondance.

12 Billatte to R, 9 October 1739, AAS-FR, Correspondance.

13 *MI* 1:56–57.

14 Boissier de Sauvages to R, 31 July 1733, BCMHN, Ms. 2624. Boissier de Sauvages later wrote a treatise on the cultivation of silkworms: P.-A. Boissier de Sauvages, *Mémoires sur l'éducation des vers à soie* (Nîmes, 1763).

15 Boisser de Sauvages to R, 2 April 1734. BCMHN, Ms. 2624.

16 *MI*, 1:63.

17 Ibid., 1:64.

18 Ibid., 1:38.

19 On representations of insect anatomy in the seventeenth century, Cobb, "Malpighi, Swammerdam and the Colourful Silkworm"; Bertoloni Meli, *Mechanism, Experiment, Disease*, 175–207. Plate 5 in *MI*, vol. 1, was copied from an illustration in Malpighi's 1669 work *Dissertatio epistolica de bombyce*. The silkworm is smaller than the caterpillars Réaumur dissected; he did not have the fine anatomical skills of his predecessor. For R's comment on Malpighi, *MI*, 1:129.

20 *MI*, 1:149.

21 Ibid., 1:149–50.

22 Ibid., 1:152.

23 Ibid., 2:11: "Que sçait-on si ce que nous est inutile aujourdhui ne deviendra pas utile quelque jour?

24 Ibid., 1:360.

25 Ibid., quotations on 1:350–51.

26 Ibid., 1:351. R's term is "*le faux merveilleux*."

27 Ibid., 1:363.

28 Ibid., 1:404–5. These experiments were done with Nollet (see chapter 3).

29 Ibid., 1:383.

30 Ibid., 1:414.

31 Ibid., 1:351.

32 Ibid., 2:xv–xvi (preface): "A mesure qu'on accordera plus d'attention aux insectes, on fera des observations qui m'ont échappé: celles meme que je rapporte sont quelquefois imparfaites; il m'arrive quelquefois de parler d'une chenille dont je n'ai pas encore eu le papillon, & de parler d'un papillon dont la chenille ne m'est pas encore connue: c'est avertir les observateurs de ce qui reste à faire, & c'est les inviter à profiter des occasions qui pourront leur faire voir en entier ce que je n'ai vu qu'à moitié."

33 Ibid., 1:387–88: "J'avois une très-grande table toute couverte de ces chenilles; aussi ne se passoit-il gueres de quarts d'heure où je n'en pusse surprendre quelqu'une dans le fort de l'operation."

34 Ibid., 1:390.

35 Ibid., 1:415.

36 "Douxième mémoir: De la construction des coques, de formes arrondies, soit de pure soye, soit de soye & poils, où differentes especes de chenilles se metamorphosent en crisalides." Ibid., 1:487–534.

37 Ibid., 1:523.

38 Ibid., 1:542.

39 Bazin to R, 12 January 1735, AAS, dossier Bazin. "Vous avez besoin d'amis qui prennent chaudement votre parti, et qui mettent souvent leur honneur en jeu pour certifier que ce n'est point le Roman de la Nature que vous avéz pretendu faire, mais les véritables faits et gestes des insectes."

40 Bazin to R, 12 January 1735, AAS, dossier Bazin. Bazin is commenting on *MI*, vol. 1, Mémoire XIV: "On the Transformation of Chrysalises into Butterflies."

41 Bazin to R, 3 June 1735, AAS, dossier Bazin. Bazin noticed details about structures in the caterpillar's mouth that Réaumur had not seen, and that Simonneau's illustrations did not represent accurately.

42 Bazin to R, 4 July 1735, AAS, dossier Bazin.

43 Bazin to R, 15 August 1736, AAS, dossier Bazin: "T. 1, P. 641 de vos mems. vous dites Monsieur 'devenu parfait papillon, il n y a plus sur les anneaux de son corps de stigmates visibles.' En maniant un papillon femelle du ver de soye, je fus surpris de voir malgré le poil dont il est chargé, des stigmates bien visibles sur les anneaux, comme ils estoient sur la crisalide." Note that *papillon* refers to both butterflies and moths.

44 Bazin to R, 3 June 1735, AAS, dossier Bazin: "I am making for myself and for my own private use a very complete table [extract] of your book. When there is a second edition, if you think it appropriate to have one, which would be in effect a very good idea, I will give you mine. You will find it all done and will only have to correct it."

45 Bazin to R, 21 June 1736, AAS, dossier Bazin.

46 Bazin to R, 15 October 1736; he sent the table of the first two volumes with Bazin to R, 28 June 1737. Both in AAS, dossier Bazin. Fair copies in Bazin's hand of the indexes for subsequent volumes are still in Réaumur's papers in the Academy of Sciences. AAS-FR, dossier 34.

47 Bazin to R, 12 January1735, AAS, dossier Bazin. There was no revised or abridged edition.

48 Bazin to R, 3 June 1735, AAS, dossier Bazin.

49 *MI*, 2:xiii.

50 Ibid., 2:xiii–xiv.

51 Bazin to R, 22 December 1738, AAS, dossier Bazin.

52 Baux to R, 15 June 1742, AAS, dossier Baux. Baux's phrase is "une histoire si belle."

53 Bonnet, *Mémoires autobiographiques*, 48–49.

54 On the seventeenth-century microscopists, see Ruestow, *Microscope in the Dutch Republic*; Wilson, *Invisible World*; Bertoloni Meli, *Mechanism, Experiment, Disease*. On early-eighteenth-century microscopical observations, especially those of the Frenchman Louis Joblot, see Ratcliff, *Quest for the Invisible*.

55 *MI*, 1:30: "Ils avoient crû les pouvoir faire naitre de la pourriture de corps de differentes especes. . . . Malgré le ridicule qu'il y a à faire naitre une Mouche à miel de la chair pourrie d'un veau ou de celle d'un boeuf, les Guespes & les Bourdons de celle d'un cheval pourri, les Scarabés de celle des ânes; à faire naitre une infinité d'autres insectes, les uns de fromage, les autres de plantes, & les autres même de la boue, il a fallu bien des observations & bien des raisonnements avant que de détruire des sentiments si absurdes."

56 Ibid.

57 *MI* 2:xxxix–xl.

58 *Mémoires pour servir à l'histoire des sciences et des arts* (known as *Journal de Trevoux)*, June 1735, 1125. Reviews in this journal were unsigned. On the use of microscopical observations to counter spontaneous generation, and the fluctuating role of the Jesuits and the *Journal de Trevoux* in the controversy, see Ratcliff, *Quest for the Invisible*, chap. 2, and pp. 124–29. An adept microscopist and committed spontaneist, Bonanni was Kircher's successor as curator of the Collegio Romano museum. Bonanni lived until 1725, but his relevant publications were in 1680s and 1690s. See Findlen, "Science, History, and Erudition."

59 *Journal de Trevoux*, June 1735, 1120. The Dutch journal *Bibliothèque françoise* applauded Réaumur's attack on "superstition." Review of *MI*, vol. 2, in *Bibliothèque françoise* 25, no. 2 (1737): 342–64, on 361 (cited by Ratcliff, *Quest for the Invisible*, 126).

60 R to Louis Bourguet, 9 December 1735, Bibliothèque publique et universitaire, Neuchâtel, Ms. 1278.

61 *MI*, 2:xvii: "I would not have suspected that [this opinion] would find defenders in France, and especially in such a celebrated society of savants as that working on the *Journal de Trevoux*."

62 Ibid., 2:xvi.

63 Ibid., 2:xxiv.

64 Ibid., 2:xl.

65 Ibid., 2:xxvi.

66 Ibid., 2:xxix.

67 Ibid., 1:xxviii.

68 Ratcliff, *Quest for the Invisible*, esp. 126–27.

69 *MI*, 1:352: "I would have been surprised that such an idea was adopted by a celebrated metaphysician, whose genius was as clear as it was sublime, if I did not know that he could sometimes be ruled by his imagination, when he was not sufficiently on guard against it." Réaumur had studied mathematics with Malebranche as a young man and was on good terms with him until his death in 1715. The Jesuits defended Malebranche even though he was an Oratorian and not one of their own (as Kircher had been).

70 *Journal de Trevoux*, June 1735, 1242: "Il n'y a point là de resurrection, il n'y a rien meme de miraculeux, puisque ces expressions révoltent: mais il y a presque

quelque chose de superieur au miracle, quelque chose au moins de bien propre à nous donner une idée d'une puissance capable d'opérer les plus grands miracles & le miracle en particulier de notre resurrection."

71 *MI*, 3:xxxvii–xxxvix.

72 Bazin to R, 18 June 1736, AAS, dossier Bazin.

73 Ibid.

74 Anon. to R, BCMHN, Ms. 2624, n.d. [1736]. This reader went on to point out other mistakes by Jesuit authors, where their prejudice in favor of Aristotle and Kircher betrayed them.

Chapter Five

1 *MI*, 6:546–47.

2 Bazin to R, 28 August 1737, AAS, dossier Bazin.

3 Bazin to R, 24 July 1739, AAS, dossier Bazin.

4 *MI*, 2:65. R first mentions his trick of pinning large numbers of different kinds of chrysalises to "a large section" of his tapestry in a *mémoire* on the mating behaviors of different species of butterflies and moths. *MI*, 1:608.

5 Ibid., 2:66.

6 Ibid., 3:282: "It is more pleasing for us to know the little maneuvers of insects found in our gardens than those of insects from the Indies that we will never see [alive]."

7 C.-J. Geoffroy, "Observations sur les vessies qui viennent aux ormes, & sur une sorte d'excroissance à peu-près pareille qui nous est apportée de la Chine," *MAS* 1724, 320–26.

8 *MI*, 3:289.

9 Ibid., 3:290–91.

10 Ibid., 3:326.

11 Ibid., 3:290.

12 Ibid., 3:327.

13 Ibid., 3:329.

14 Ibid., 3:329–30.

15 Bonnet to R, 4 July 1738, AAS, Dossier Bonnet. Bonnet's autobiography, written many years later, recounts his early enthusiasm for natural history, starting with Pluche. He knew at least the first volume by Réaumur when he first wrote to Paris in July 1738. Bonnet, *Mémoires autobiographiques*.

16 Bonnet often sent boxes of specimens with his letters; e.g., Bonnet to R, 27 July 1739, AAS, dossier Bonnet.

17 Charles Bonnet, "Préface," in *Traité d'insectologie*, xxxi.

18 Bonnet, *Traité d'insectologie*, xxv.

19 Bonnet to R, 13 July 1740, AAS, Dossier Bonnet.

20 Ibid.

21 On Bonnet's attachment to his aphid, see Daston, "Attention and the Values of Nature," 106 (and, more generally, the section on "Creature Love"). See also Dawson, *Nature's Enigma*, 78–80.

22 R to Bonnet, 5 August 1740, BG, Ms. Bonnet 42.

23 Bazin continued with his experiments on aphids the following summer; see Bazin to R, 14 June 1741, AAS, dossier Bazin. Lyonet reported his results in Lyonet to R, 28 March 1742, AAS-FR, Correspondance.

24 AAS p-v, 23 July 1740. Many of R's correspondents had to wait for years for this form of recognition.

25 AAS p-v, 26 July 1740, for Réaumur's presentation. Bonnet refers to R's discussion in *MI*, 3:334, in Bonnet to R, 23 November 1740, AAS, dossier Bonnet.

26 Bonnet to R, 23 November 1740, AAS, dossier Bonnet.

27 *MI*, 2:xxxix.

28 Bonnet read Swammerdam and Malpighi with Cramer. Bonnet, *Mémoires autobiographiques*, 55–56. Cramer and Calandrini edited a review jounal, *Bibliothèque italique*, with French extracts of books published in Italy. On Trembley's life on the Bentinck estate, see Dawson, *Nature's Enigma,* 85–88.

29 "Il y a déjà deux ans que le mémoire de Mr. de Réaumur m'avoit donné l'envie de la faire." Trembley to Bonnet, 4 October 1740, in Dawson, *Nature's Enigma*, 193.

30 Inspired by Réaumur's experiment with the effects of temperature on insects in different stages, Trembley used the occasion of the extremely cold winter of 1740 to study the effect of cold on insects. Trembley to Bonnet, 4 October 1740, in Dawson, *Nature's Enigma*, 193; Trembley to R, 15 December 1740, *Corr. inédite*, 10.

31 Trembley to R, 15 December 1740, *Corr. inédite*, 15.

32 Van Seters, *Pierre Lyonet*, 16, 65–66. Lyonet studied drawing when he came to The Hague as a young lawyer, and was a friend of the Dutch artist Hendrik Van Limborch, who painted his portrait in 1742 (now in the Museum of Natural History in Leiden). Lyonet had strong opinions about the quality of natural history illustration; he may have been motivated initially by the crudeness of the plates in Pluche's *Spectacle de la nature.*

33 Réaumur was confused about the stimulus for Lyonet's aphid observations; he assumed that Trembley had told Lyonet about Bonnet's results. Lyonet insisted that he had been inspired only by reading about Réaumur's inconclusive experiments in *MI*, vol. 3. See Lyonet to R, 22 February 1743, HL, Ms. Fr. 99. For the proliferation of aphids in the summer of 1740, Trembley to Bonnet, 4 October 1740, BG, Ms. Bonnet 24.

34 Trembley first mentioned Lyonet, "un de mes amis, grand observateur," to Bonnet, 4 October 1740; in his next letter (27 January 1741, BG, Ms. Bonnet 24), he gave a rather more detailed account of what Lyonet had seen.

35 All of Lyonet's notes on aphids may not be extant. Several pages recording

observations from the 1740 season, including one sheet of drawings of aphids, are in the collection of Lyonet manuscripts in the Bibliothèque municipale de Mons, Ms. 512. Some of Lyonet's later experiments are documented only in his letters to Réaumur.

36 Lyonet, "Puceron verd du saule vivipare," Bibliothèque municipale de Mons, Ms. 512.

37 Lyonet, "Suite du puceron No. 2," Bibliothèque municipale de Mons, Ms. 512.

38 Bonnet had also seen his aphids mating in the same autumn.

39 Lyonet, "Puceron sociable ovipare," Bibliothèque municipale de Mons, Ms. 512.

40 Ibid.

41 Trembley to R, 9 February 1741, *Corr. inédite*, 20.

42 Trembley to Bonnet, 27 January 1741, in Dawson, *Nature's Enigma*, 197: "Nous comptons nous y mettre dès le commencement du printems, et nous vous invitons à faire les memes experiences de votre coté."

43 Van Seters explains the complicated business dealings among the printers and publishers. Van Seters, *Pierre Lyonet*, 87.

44 Lyonet, note to Lesser, *Théologie des insectes*, 50.

45 Lyonet, note to Lesser, *Théologie des insectes*, 50–51.

46 Réaumur used the term "foetus avortés" in R to Lyonet, 17 April 1742, BG, Ms. Trembley 5.

47 Lyonet to R , 28 March 1742, AAS-FR, Correspondance.

48 "Que dans la meme genre d'insectes il y en ait des especes vivipares, et d'autres ovipares, nous ne devons plus en etre etonnés. Les mouches a deux aisles nous en donnent des exemples. Mais il doit paroitre bien surprenant que dans la meme espece il y ait des individus vivipares et d'autres ovipares, et cela selon la saison dans laquelle ils travaillent a perpetuer leur espece." R to Lyonet, 17 April 1742, BG, Ms. Trembley 5.

49 Lyonet to R, 11 May 1742, BG, Ms. Trembley 5.

50 Ibid.

51 *MI* 6:534–35. Bonnet used these in his own account (*Traité d'insectologie*), but it was not published until 1745.

52 *MI*, 6:538. ". . . ils ont vû & revû des pucerons de différentes especes qui ont mis des petits au jour."

53 Ibid., 6:541: "Je mériterois cependant des reproches si je n'avois pas cherché à m'assurer par mes propres yeux, d'une vérité qui avoit été démontrée par des expériences du soin desquelles d'autres avoient bien voulu se charger *à mon instigation*" (my emphasis).

54 Ibid., 6:542–43.

55 Ibid., 6:543–44.

56 Ibid., 6:524.

57 Ibid., 6:553.

58 Ibid., 6:560: "petits corps oblongs;" "des especes d'oeufs." Réaumur reported Bonnet's observation of an aphid emitting these bodies and arranging them on a branch, much as a butterfly lays its eggs.

59 Ibid., 6:559.

60 Lyonet to R, 11 April 1743, HL, Ms. Fr. 99.

61 Trembley to Bonnet, 26 July 1740, BG, Ms. Bonnet 24.

62 Trembley to R, 15 December 1740, *Corr. Inédite*, 11.

63 On Trembley's discovery, see Dawson, *Nature's Enigma*, 85–136. See also Ratcliff, "Trembley's Strategy of Generosity."

64 He originally cut the creatures to settle the question of whether they were plants or animals. Trembley to R, February 1741, *Corr. inédite*, 24.

65 R, "Sur les diverses reproductions qui se font dans les écrevisses, les omars, les crabes, etc.," *MAS* (1712), 223–41.

66 Trembley to R, 15 December 1740, *Corr. inédite*, 15.

67 Trembley's microscope is depicted in Needham, *Nouvelles découvertes*, plate 7. See also Ratcliff, *Quest for the Invisible*, 82–84.

68 Trembley to R, 16 March 1741, *Corr. Inédite*, 56.

69 Ibid., 57–58.

70 Ibid., 58.

71 Ibid., 53. See Ratcliff, *Quest for the Invisible*, 106–7, 110, on Trembley's experimenting with techniques for sending live polyps.

72 AAS p-v, 1, 8, and 22 March 1741.

73 The creature is now known as a hydra; I use the contemporary term "polyp" throughout.

74 R to Trembley, 25 March 1741, *Corr. inédite*, 65.

75 See J. J. Dortous de Mairan, "Animaux coupés & partagés en plusieurs parties . . . ," in *HAS* 1741, 33–35.

76 Trembley to Bonnet, 8 February 1743, BG, Ms. Bonnet 24; see also Johan-Friedrich Gronovius, "Concerning a Water Insect, which, being cut into several Pieces, becomes so many perfect Animals," *Philosophical Transactions* 42, no. 466 (1742): 218–20.

77 Trembley to R, 8 June 1741, *Corr. inédite*, 78–79.

78 Trembley to Bonnet, 11 December 1742, BG, Ms. Bonnet 24.

79 Bonnet to Trembley, 27 December 1742, in Dawson, *Nature's Enigma*, 218.

80 Trembley to Bonnet, 8 February 1743, in Dawson, *Nature's Enigma*, 219. It is not clear why he didn't send polyps to Bonnet in Geneva. See Ratcliff, *Quest for the Invisible*, 109n33, for citation to letter from Trembley to Martin Folkes, president of the Royal Society of London, on supplying polyps. See also Ratcliff, "Trembley's Strategy of Generosity."

81 Lyonet to R, 11 May 1742, BG, Ms. Trembley 5.

82 Ratcliff, "Trembley's Strategy of Generosity."

83 Trembley to R, 14 March 1743, *Corr. inédite*, 159–60: "Le pays des polypes est

si vaste, et je souhaite si fort qu'il soit connu, que je voudrais que des centaines de personnes fussent déjà occupées à le parcourir. J'espere que la preface de votre sixieme volume aura engagé plusieurs naturalistes à se mettre à chercher et à observer des polypes, et que dans quelques années nous connaitrons encore plus de merveilles que celles que nous connaissons à présent."

84 Réaumur recommended to Father Mazzoleni that he look for worms in order to see regeneration. R to Mazzoleni, 27 November 1741, Biblioteca laurenziana, Ms. Ashburnham 1522.

85 Réaumur mentioned Guettard's work and his place in the household in R to Louis Bourguet, 1740, Bibliothèque de Neuchâtel, Ms. 1278. Guettard was in Charenton for the observations of mating frogs in 1740; he apparently lived in the household until around 1745. See *Almanach royal* for 1743 and 1744.

86 Guettard to R, 27 September 1741, AAS, dossier Guettard. For Guettard's later reflections on the polyp and his own involvement in studies of regeneration, see Guettard, *Mémoires*, vols. 2 and 4.

87 Girard de Villars to R, 14 November 1741, AAS, dossier Girard de Villars.

88 Girard de Villars to R, 20 January 1742, AAS, dossier Girard de Villars. On Louis Richard, see Torlais, *Réaumur*, 336n1.

89 Girard de Villars to R, 4 May 1742, Bibliothèque de La Rochelle, Ms. 664, ff. 143–45. The friend's wife was an admirer of Réaumur, and a beekeeper; Villars reported that she was in the midst of reading the insect volumes.

90 Girard de Villars to R, 4 July 1742, AAS, dossier Girard de Villars.

91 BCMHN, Ms. 972, f. 141—R's notes on drawings of étoiles de mer: "Une obs. que fait hier M. Guettard, et celle que j'ai faite aujourd'hui, nous ont convaincu l'un et l'autre que chacune de ces petites etoilles est un animal. . . . Quelques étoilles ont 8 rayons, d'autres six ou sept, d'autres moins, et Mlle. du Moutier en a trouvée une qui n'en avoit que trois."

92 *MI*, 6:lxvi.

93 Ibid., 6:l–li.

94 Ibid., 6:liv–lv.

95 Lyonet, note to Lesser, *Théologie des insectes* 1:74.

96 Trembley, *Mémoires pour servir à l'histoire des polypes*. The first five plates were engraved by the very accomplished artist Jan van der Schley; the rest were engraved by Lyonet. A smaller-format Paris edition was printed shortly thereafter, with poor-quality copies of the plates. On the production of the plates, see van Seters, *Pierre Lyonet*.

97 Bazin, *Histoire naturelle des abeilles* (1743); English translation, 1744.

98 Bazin, *Lettre d'Eugène à Clarice au sujet des animaux appellés Polypes* (Strasbourg, 1745); Bazin, *Abrégé de l'histoire naturelle des insectes*. On Bazin's dialogues, see Olivier, "Bazin's 'True Novel' of Nature."

99 Bazin, *Abrégé de l'histoire naturelle des insectes*, 2:xiv.

100 Ibid., 2:185.
101 Ibid., 2:194.
102 Ibid., 189.

Chapter Six

1 R, "De l'utilité des cabinets d'histoire naturelle et de l'objet de cet ouvrage qui est d'assurer la durée des collections qui doivent entrer dans ces cabinets," autograph ms., AAS-FR, carton 1. This was intended as the first section of a book, never published, on how to build and maintain a collection.
2 *MI* 4:xxvi–xxvii.
3 For contemporary lists of Paris collections, see Gersaint, *Catalogue raisonné*, 30–45; Dézallier d'Argenville, *Histoire naturelle éclaircie*, 198–210; Laissius, "Cabinets d'histoire naturelle," identifies more than two hundred proprietors of natural history collections. On collections more generally, Pomian, *Collectors and Curiosities*.
4 Dézallier d'Argenville, *Histoire naturelle éclaircie* (1742), 196; Daubenton repeats this prescription in "Description du Cabinet du Roy," 91-2. See also Dietz, "Mobile Objects"; Bleichmar, "Learning to Look."
5 When Buffon took over the direction of the Jardin du roi in 1739, he brought in his protégé Daubenton, displacing Réaumur's friend Bernard de Jussieu as curator of the collection. See Spary, *Utopia's Garden*, 39-41.
6 For a Renaissance collection that served also as a laboratory and a place for conversation and dispute, see the rich portrait of Ulisse Aldrovandi's museum in Bologna in Findlen, *Possessing Nature*.
7 Lignac, *Lettres à un américain,* 81: "Il le dit volontiers à tous ceux qui viennent le visiter, & quand il ne le diroit pas, des étiquettes attachées à chaque pièce, apprendroient assez à qui elles sont dues."
8 One example: R to Crousaz, 30 October 1719, Bibliothèque universitaire de Lausanne, Fonds Crousaz, box XIV: "J'ai deja dans mon cabinet des pirites a faces regulieres, pareilles a celles que vous avez adressés a Mr. l'Abbé Bignon; j'y placerai cependant les votres, quand ce ne seroit que pour avoir le plaisir de trouver de temps en temps un etiquette qui portera un nom que je respecte si fort." Réaumur mentions that contributors' names are put on labels: R to Bentinck, 31 July 1748, in *Corr. inédite*, 308.
9 Lignac, *Lettres à un américain*, 2 (pt. 4): 82.
10 R's notes on experiments, dated 1729, AAS-FR, dossier 19.
11 Gersaint, *Catalogue raisonné*, 32–33. Gersaint's term for natural philosopher was "*physicien habile*." Gersaint was especially interested in shells for their commercial value. On the fashion for shell collecting in France, see Spary, "Scientific Symmetries"; Dietz, "Mobile Objects." On Gersaint, see Glorieux, *A l'enseigne de Gersaint*.

12 Gersaint, *Catalogue raisonné*, 32.

13 All the documents produced by the *enquête* have been published, with introduction and extensive notes, in Demeulenaere-Douyère and Sturdy, *Enquête du Régent*.

14 "Projet de lettre circulaire de Mgr le duc d'Orléans à MM. les Intendans," n.d. [January 1716], in Demeulenaere-Douyère and Sturdy, *Enquête du Régent*, 77.

15 For a proposal to establish a study collection for the institution, see Etienne-François Geoffroy to Bignon, 1716, in Demeulenaere-Douyère and Sturdy, *Enquête du Régent*, 81–83.

16 R, "Remarques sur les coquilles fossiles de quelques cantons de la Touraine, et sur les utilités qu'on en tire," *MAS* (1720), 400–416, on 401. At least six of Réaumur's academic papers drew on material collected in the *enquête*, which also sparked his extensive work on iron and steel in the 1720s. See also R to Bourguet, 8 April 1738 (Bibliothèque publique et universitaire, Neufchâtel, Ms. 1278), where he comments on the extent of the fossil collection and the interest of the late regent in all this.

17 R, "Second mémoire sur la porcelaine, ou suite des principes qui doivent conduire dans la composition des porcelaines de différens genres," *MAS* (1729), 329.

18 Lignac mentions the Potosi gold, "ce morceau unique d'or de M. d'Osenbrai." Lelarge de Lignac, *Lettres à un américain*, 2 (pt. 4): 39. D'Argenville, who had seen an impressive lump of Potosi gold in Pajot d'Onsenbray's collection in the early 1740s (perhaps the one given to Réaumur), also mentioned the specimens from the Harz, as well as a large cabinet of fossils. Dézallier d'Argenville, *Histoire naturelle éclaircie,* 205–6.

19 Réaumur may well have brought back Swedish samples himself; he visited the Royal College of Mines in Stockholm in 1724. See Swedenborg to Benzelius, 20 August 1724, in Tafel, *Documents* 1:340. I thank Hjalmar Fors for this reference.

20 D'Argenville, *Histoire naturelle*, 206.

21 R to Pierre Baux, 1744, Bibliothèque municipale de Nîmes, Ms.

22 The neighborhood was also home to several convents and a hospital, large establishments that needed more space than they could find in town.

23 "Je m'eloigne de 4 a 5 minutes du centre de Paris, de plus que je n'en suis eloigné actuellement, pour habiter un maison qui a de tres grands et tres beaux jardins, et de tels qu'on n'en trouve point dans le coeur de la ville. Je me trouverai a la ville et a la campagne." R to Bourguet, 10 July 1740, Bibliothèque publique et universitaire, Neufchâtel, Ms. 1278.

24 Blondel, *Architecture françoise*, 1:135. The house had been designed by the architect Dulin in the first decade of the eighteenth century for a wealthy official of the Paris *parlement*, M. Dunoyer; it was known in the Regency as the site of various scandalous assignations. Réaumur leased the house with its ancil-

lary buildings, grounds, and furnishings from the niece of Dunoyer, Mme. de Winterfeldt. "Bail," 6 April 1740, AN-MC, ET/XCIV/220.

25 Blondel, *Architecture françoise*, 1:135–36.

26 Ibid., 1:135.

27 Trembley to Bonnet, 22 December 1752, quoted in *Corr. inédite*, 367n2 (original in BG, Ms. Bonnet 15). Trembley mentioned the presence at these dinners of Nollet, Duhamel de Monceau, and Bernard de Jussieu, showing the continuity of Réaumur's social ties across the decades.

28 Lignac, *Lettres à un américain*, 2 (pt. 4): 30.

29 "Inventaire après déces de M. Ferchault de Réaumur," November 1757, AN-MC, ET XCIX/534. Jean-Etienne Guettard made a catalog of some parts of the collection when he was working there in the 1740s, including information about the provenance of many specimens, probably taken from their labels. "Inventaire partiel du cabinet de M. de Réaumur," BCMHN, Ms. 1929/iii, ff. 80–183.

30 Brisson, assistant to Réaumur from 1749 to 1757, based his classification of quadrupeds on this collection; his book is another source for the range of the collection of mammals. See Brisson, *Règne animal*. In her discussion of the overlapping worlds of connoisseurs of art and naturalia, Bleichmar points out the continuing presence in eighteenth-century collections of typical Renaissance curiosities like dried alligators hanging from ceilings. Bleichmar, "Learning to Look."

31 Lignac, *Lettres à un américain*, 2 (pt. 4): 35. The notaries valued the contents of these three rooms of birds, all in sealed glass cases, at more than 1,800 *livres*.

32 These birds are specified in the inventory; Lignac mentions a few of them, like the ostrich. Ibid.

33 Brisson, *Ornithologie*, included some donor names in his descriptions.

34 Séguier sent several chests of fossils he had collected in Italy. See, e.g., Séguier to R, 22 February 1745; 6 February 1748; 4 September 1748. AAS, dossier Séguier. Louis Bourguet had sent several boxes of fossils from Switzerland, mentioned in R to Bourguet, 11 December 1738, Bibliothèque publique et universitaire, Neufchâtel, Ms. 1278.

35 Inventaire, AN-MC, ET XCIX/534.

36 Ibid. This piece of furniture was also mentioned by Dézallier d'Argenville, *Histoire naturelle éclaircie*, 206.

37 The marquis de Caumont went to considerable trouble to acquire a flamingo for Réaumur; this became "un grand ornement de mon cabinet." R to Caumont, 23 February 1750, BG, Fonds Trembley 5.

38 Guettard, "Inventaire partiel du cabinet de M. de Réaumur," BCMHN, Ms. 1929/iii.

39 Guettard, "Inventaire partiel du cabinet de M. de Réaumur," BCMHN, Ms. 1929/ iii. Pelican sent from Savoy by Nollet, 1749. AAS-FR, dossier 20.

40 R to Baux, 12 April 1745, Bibliothèque municipale de Nîmes.

41 R to Crousaz, 30 October 1719, Bibliothèque universitaire de Lausanne, Fonds Crousaz, box XIV.

42 R, "Moyens d'empêcher l'évaporation des liqueurs spiritueuses, dans lesquelles on veut conserver les productions de la Nature de différens genres," *MAS* (1746), 516–38.

43 R to Mazzoleni, May 30, 1744, Biblioteca laurenziana, Ms. Ashburnham 1522.

44 The phrase comes from R to Trembley, 21 May 1744, *Corr. inédite*, 187.

45 [R], *Différens moyens*.

46 R to Trembley, 21 May 1744, *Corr. inédite*, 187.

47 Ibid. R. told Trembley that he had started seriously collecting birds in autumn 1743.

48 R to Séguier, 16 November 1744, in *Lettres inédites*, 24–25. He told Séguier that it only takes fifteen minutes to learn how to prepare the birds. The only other large collection of birds known to him was in Berlin in the cabinets of Frisch, who had only"217 birds in total."

49 R to Trembley, 9 July 1744, *Corr. inédite*, 192. He also told Séguier, Artur, and Mazzoleni about the device. The blacksmith's apparatus was known as a *travail de maréchal*. The "machine" is described in the pamphlet, and in English translation as "Divers Means for preserving from Corruption dead Birds, intended to be sent to remote Countries, so that they may arrive there in a good condition," trans. P. H. Zollman, *Philosophical Transactions* (1748), 304–20.

50 Réaumur described several versions of the drying frame to various correspondents. E.g., R to Artur, 25 March 1749, in Chaia, "Sur une correspondance," 125; *Differens moyens*, 3.

51 R to Trembley, 9 July 1744, *Corr. inédite*, 192. On embalming processes used to ship naturalia around the world in an earlier period, see Cook, "Time's Bodies."

52 R to Caumont, 5 January 1745, BGE. In 1746, when he presented his public lecture to the academy on methods for preventing evaporation from specimen jars, Réaumur noted, "If the seventh volume on insects has not yet appeared, it is not that I think these little animals can offer us no further marvels, or that I have become indifferent to those they have yet to show us; but it seemed to me that a work aimed at advancing different parts of natural history, some of the most interesting of which have been too neglected, for want of knowing how to make up the curious collections that they might provide to us, that such a useful work (useful even for collections of insects) had sufficient justification to be published before the continuation of the other." *MAS* (1746), 484.

53 R to Séguier, 16 November 1744, in *Lettres inédites*, 25.

54 Most of a draft manuscript (about three hundred sheets) on the preservation of collections is in AAS-FR, carton 1. The existence of a nearly complete

manuscript was known to Réaumur's contemporaries (e.g., Bonnet). A chapter on insects that prey on collections was published posthumously as part of the "Natural History of Beetles" in Caullery, *Papiers laissés par de Réaumur*, Mem. 7 ("Des différentes espéces d'insectes contre lesquelles on a à défendre les collections d'oiseaux et toutes celles du règne animal").

55 Réaumur mentions long transit times, *Différens moyens*, 2. Although several sources (e.g., Torlais, *Réaumur*) state that the pamphlet is not known, several copies survive, including those in HL and BCMHN. For mentions of the pamphlet, see R to Trembley, 28 January 1747, *Corr. inédite*, 289; R to Artur, 1 May 1746 and 9 February 1747, in Chaia, "Sur une correspondance." An English translation was read to the Royal Society of London, as noted above; [R], "Divers Means."

56 R to Trembley, 13 January 1746, *Corr. inédite*, 250–51.

57 Ibid. Trembley's letters from this period do not survive. Réaumur thanks him for the translation and for engaging Bentinck's support in R to Trembley, 13 March 1746 and 5 May 1746, *Corr. inédite*.

58 [R], *Différens moyens*. For English translation, see n40 above.

59 Adanson became an avid collector himself; see his *Histoire naturelle du Sénégal*. His personal copy of *Différens moyens* is held in BCMHN. Brisson, *Ornithologie*, records numerous African birds sent by Adanson to Réaumur. There were two versions of the pamphlet, in 1745 and 1747; Séguier had the first version, *Moyen facile de conserver les oyseaux* (Séguier to R, 7 November 1745, AAS, dossier Séguier.) See also Adanson to R, 15 August 1749, Bibliothèque nationale de France, Ms. n.a.f. 5151.

60 [R], *Différens moyens*, 1.

61 Ibid., 3.

62 Poivre to R, 19 March 1757, AAS dossier Poivre.

63 On Séguier, see Roche, *Républicains des lettres*, 263–80; Chapron, "Echanges savants." On the Séguier-Réaumur correspondance, Torlais, *Réaumur*, 323–26. Séguier's letters to Pierre Baux in Nîmes chronicle his activities, and the hundreds of books he bought, in Paris. Cordier and Pugnière, *Seguier, Baux Lettres*.

64 Réaumur recalls the visit; R to Sèguier, 25 April 1743, *Lettres inédites*, 15. For fossils, Séguier to R,13 December 1744, AAS, dossier Séguier; R to Séguier, 10 January 1744, in *Lettres inédites*, 20. Séguier also sent sent fossils to Mme de Verteillac, another Parisian collector, via Réaumur; see R to Séguier, 25 May 1748, in *Lettres inédites*, 71.

65 Séguier made a list of local birds, identified from reference works by Aldrovandi and Willughby, and asked Réaumur to indicate which of these he wanted. Séguier to R, 22 February 1745, AAS, dossier Séguier.

66 Séguier to R, 7 November 1745, AAS, dossier Séguier.

67 R to Séguier, 18 May 1746, in *Lettres inédites*, 42.

68 Séguier to R, 20 August 1746, AAS, dossier Séguier.

69 "Quoi, voilà déjà des oiseaux de Cap de Bonne Espérance arrivés! Il semble que Monsieur le comte de Bentinck n'ait eu qu'à parler." R to Trembley, 31 July 1748, *Corr. inédite*, 305. See also R to Bentinck, 31 July 1748, BL Egerton ms. 1745.

70 R to Bentinck, 6 November 1748, in *Lettres inédites*, 151.

71 R to Bentinck, 29 December 1748, BL Egerton mss. 1746; also (from Trembley's copy) in *Corr. inédite*, 314–15.

72 R's name for these geese was *anser bassanées*. R to Morton, 3 December 1748, Beinecke Library, OSB Mss. 72. In 1753, Réaumur welcomed Lord Aberdour, Morton's son, to Paris, entertaining him at dinner and undoubtedly taking him around the collection.

73 Seeing the birds from Scotland sometimes allowed Réaumur to clear up confusion about names and to see varieties subtly different from those already on his shelves. "The bird called a 'scoub' in Scotland seems to me to be the one they call *guillemot* on our Normandy coast, where it comes to lay its eggs every year in the spring; someone sent me some of its eggs which are bluish, marbled with black." R to Morton, 23 January 1750, , Beinecke Library, OSB Mss. 72.

74 R to Morton, 5 May 1749, Beinecke Library, OSB Mss. 72.

75 R to Morton, 7 Sept 1752, Beinecke Library, OSB Mss. 72.

76 In his accounting of the costs of managing the collection (in a request for future funding by the crown), Réaumur included a line item for supplies and "shipping costs for the chests and packages which arrive daily from all locations." Réaumur to unknown "monseigneur," n.d. [1755 or 1756], in Caullery, *Papiers laissés par Réaumur*, 30.

77 Artur met Réaumur through his teachers at the Jardin du roi, Bernard de Jussieu and Charles Dufay. Eventually he served on the governing council of Cayenne and ran his own cocoa and cotton plantation with slave labor. On Artur's career in Cayenne, see Polderman, "Manuscrit de Jacques François Artur"; Ronsseray, "Destin guyanais."

78 On Artur's trouble with temperature measurements, see Bourguet, "Measurable Difference." Artur was only named an official correspondent of the Academy of Sciences many years later, in 1753.

79 R to Artur, 15 March 1741, in Chaia, "Sur une correspondance," 43: "Si, Monsieur, vous estes disposé a me faire de tels presents, c'est a mon adresse et non a celle du jardin du Roy, qu'il les faudra metre, je serai charmé d'apprendre a Mr. le compte de Maurepas, a l'Académie, et au public ce que je vous deverai . . ." Artur was also in touch with Buffon, though their correspondence was perfunctory.

80 R to Artur, 15 March 1741, in Chaia, "Sur une correspondance," 42. For another example of this technique for preserving and shipping, by Réaumur's correspondent the Marquis de Caumont, see Terrall, "Following Insects Around."

81 R to Artur, 1 December 1741, in Chaia, "Sur une correspondance," 47.

82 Ibid., 47–48. Artur used tafia, the local strong liquor, in future shipments: Artur to R, 2 June 1743, AAS, dossier Artur.

83 Réaumur credited Artur with the observation of these wasps. *MI*, 6:229.

84 25 March 1749, in Chaia, "Sur une correspondance,"126. He had mentioned the feather temperature indicator earlier; R to Artur, 1 December 1741, in ibid., 49.

85 He also mentioned the possibility of stuffing carcasses with salts or aromatic powders, including alum, though "this might be rare and expensive in your colony." R to Artur, 1 May 1746, in Chaia, "Sur une correspondance," 121.

86 R to Artur, 25 March 1749, in Chaia, "Sur une correspondance," 126.

87 Ibid. We do not know whether this meeting took place as planned, or whether Artur adopted the device in the tropics.

88 R to Artur, 9 February 1747, in Chaia, "Sur une correspondance," 123. The first time this happened is mentioned in R to Artur, 1 May 1746, in ibid., 121: "Si j'en eusse été instruit a temps je les eusse reclamées en angleterre avec esperance de les ravoir, mais je crois qu'il est trop tard pour y songer."

89 R to Trembley (in London), 9 November 1747, *Corr. inédite*, 296.

90 Inventories survive for some of the shipments: e.g., Artur to R, 2 June and 14 July 1743 (the latter with appended "État de la caisse de M. Réaumur") AAS, dossier Artur. Brisson's *Ornithologie* attributes many specimens to Artur.

91 For the elephant skin, Trembley to R, n.d.; Bucknell to R, 13 July 1756, AAS-FR, Correspondance. In this respect Réaumur's correspondents were unlike the botanical network managed by André Thouin at the Jardin du roi in the next generation. Thouin traded seeds and plants with his counterparts at other gardens and collections, but he did not maintain personal relationships with these correspondents. See Spary, *Utopia's Garden*, chap. 2.

92 Brisson, *Ornithologie*. On Brisson's book, see Spary, "Codes of Passion." Guettard's inventory also preserves many names of contributors, taken from the labels. "Inventaire partiel du cabinet de M. de Réaumur," BCMHN, Ms. 1929/iii, ff. 80–183.

93 On Brisson and Martinet, see Spary, "Codes of Passion." Spary observes that "these works of natural history were not just discourses *about* the collection; they were one level of the collection itself" (113). See also Fouchy's account of Brisson in *HAS* (1759), 59, based on Brisson's preface. Brisson says in the introduction to *Règne animal* (1756) that he planned to publish "a general description of the magnificent natural history cabinet of M. de Réaumur"; this was never completed. Brisson divided the birds into twenty-six orders and 115 genera. The plates by Martinet represented more than five hundred birds; the six volumes described around fifteen hundred species or varieties.

94 Brisson, *Ornithologie*, 1:128, 123, 195. Dozens of contributors are named by Brisson throughout his six volumes, including Altman (from the Swiss Alps),

Chervain (Saint Domingue), Poivre (China, Indochina, Philippines), Gaultier (Quebec), and Godeheu de Riville (Malta).

95 On Réaumur's social circles, see Badinter, *Passions intellectuels*, 38–41.

Chapter Seven

1 See Brisson, *Ornithologie*, 1:164–75, for the species in genus Gallinaceum, all taken from "the museum of M. de Réaumur." Most of these match the varieties in Réaumur's experiment notes: black (Mozambique), *frisée*, Persian, Indian, dwarf, five-clawed, Paduan (*poule de* caux). and others. Réaumur mentioned the growing number of chickens in his cabinet as early as 1745; R to Mazzoleni, 26 August 1745, Biblioteca Laurenziana, Ms. Ashburnham 1522.

2 Blondel specifies the location of the *basse-cours*, between the house and the street entrance. Blondel, *Architecture françoise*, 135.

3 *Art de faire*, 1:81.

4 R to Lyonet, 10 May 1747. BG, Ms. Trembley 5: "Ma collection d'oiseaux a cru plus rapidement que je ne l'avois esperé. Elle m'a fourni des observations, elle m'a fait naitre des vues, elle m'a engagé à des experiences et j'ai été conduit à avoir des regrets de ce que l'ornithologie, une de plus curieuses et des plus interessantes parties de l'histoire naturelle, étoit encore si imparfaite et si negligée." Réaumur was explaining how ornithology had sidetracked the completion of his final volume on insects.

5 Reference to these investigations can be found throughout *Art de faire*, especially in the final chapter, "Esquisse des amusemens philosophiques que les oiseaux d'une basse'cour ont à offrir." The digestion experiment is reported in R, "Sur la digestion des oiseaux. Second mémoire: De la manière dont elle se fait dans l'estomac des oiseaux de proie," *MAS* 1752, 461–95. Manuscript of this paper, and many other notes, in AAS-FR, dossiers 21 and 22; for Maitre Guillaume's role in chicken crosses, see "Note sur les couvées," AAS-FR, dossier 25. Réaumur left a bequest to the gardener "Guillaume" in his will. Cramer noted that Réaumur's gardener was spending a lot of time on the poultry operation in spring 1747; Cramer to Jallabert, n.d. [early spring 1747], BG, SH 242.

6 Réaumur called his method for hatching chicks a "*petit art*," the same term he used for his innovations in taxidermy.

7 Anecdote about Henri IV in *Art de faire*, 2:231.

8 R to Séguier, 27 November 1747, *Corr. inédite*, 66.

9 R to Mazzoleni, 20 March 1748, Biblioteca Laurenziana, Ms. Ashburnham 1522: "Que j'aurois de plaisir a vous y faire manger des poulets nés dans le fumier! L'experience vous confirmeroit dans le sentiment qu'un raisonnement tres juste vous a fait prendre. Le lieu ou naissent des poulets ne doit point influer sur le goust de leur chair; si ce goust peut etre rendu plus ou moins agreable, ce ne peut etre que selon la qualité des aliments dont ils auront été nourris."

10 The number of birds fluctuated. Réaumur mentions some years having fifty to sixty hens; in another place he refers to three hundred birds altogether in the yard. R, *Art de faire*, 1:15.

11 Hérissant's academic papers based on his experiments on birds living in the yard include "Observations anatomiques sur le mouvement des becs des oiseaux," *MAS* 1748, 345–86, and "Observation anatomique sur les organes de la digestion d'un oiseau appelé Coucou," *MAS* 1752, 417–23.

12 Réaumur noted in a draft manuscript on preservation methods that "the method of M. Brisson differs from that of Abbé Menon." AAS-FR, dossier 19. Menon lived with Réaumur from 1748 until his death in 1749.

13 Ludot mentioned seeing Réaumur marking eggs once when he came to visit in 1746: "Je vous trouvai un jour tenant deux oeufs sur lesquels vous mites une datte avant de me donner audience. Je ne devinai point alors ce que vous vouliez faire de ces oeufs, et n'osai vous le demander, mais votre livre me l'a appris." Ludot to R, 24 December 1749, AAS-FR, Correspondance.

14 Both quotations from Cramer to Jallabert, 17 November 1747, BG, S.H. 242, reporting on public session of the Academy of Sciences. For a description of the butter thermometer, see R, *Art de faire,* 1: 137–39.

15 *Art de faire*, 1:95.

16 Ibid., 2:51.

17 Ibid., 2:97.

18 Ibid., 1:97.

19 "Mr. de Réaumur a fait voir la facilité & la sureté de cette operation: il a peint avec graces la singularité du spectacle que donnent plusieurs centaines de poulets, les uns prets à éclore, les autres deja éclos." Cramer to Jallabert, 17 November 1747, BG, S.H. 242.

20 *Art de faire*, 1:94–95.

21 "J'ai peine à arreter ici, et surtout à la cour, l'impatience de ceux qui veulent en faire usage avant que d'etre assez instruits, et surtout celle de nos princes et nos princesses. Le curé de Saint Sulpice en a déja un établissement qui réussit bien, à sa communauté de l'Enfant-Jesus. Les personnes qui y président sont venues prendre suffisamment de leçons de mon jardinier qui est le grand maitre de ce nouvel art." R to Séguier, 25 May 1748, in *Corr. inédite*, 72.

22 *Art de faire*, 1:110. The incubator under the overhanging roof appears in the vignette heading the third *Mémoire*.

23 Ibid., 1:111: "Le lieu où ont été mis d'abord les fours à la ménagerie de Versailles, est à peu pres dans le cas de ma remise; . . ."

24 Ibid.

25 Ibid., 1:271. On La Muette, see Martin, *Dairy Queens*, 122–24; for Nollet's quarters, Torlais, *Abbé Nollet*, 204.

26 *Art de faire*, 2:203.

27 Ibid., 2:205–6. On the cultivation of birds and other animals, including

chickens, at Versailles, see Martin, *Dairy Queens*, 124-7. In 1748, the New Menagerie, dedicated especially to farming and horticulture, was under construction in the park near the Trianon; the poultry yard was a favorite with Louis XV, who liked to show visitors around the new menagerie "filled with chickens that he loved," according to the Duc de Croy. Martin, *Dairy Queens*, 125. Réaumur reported on the different kinds of birds hatched in the Versailles menagerie in R to Mazzoleni, 30 August 1748, Biblioteca Laurenziana, Ms. Ashburnham 1522: "Outre des poulets on a fait eclore par les mesmes moyens, a la menagerie de Versailles des perdereaux, de faisandeaux, des pannaux et des oiseaux de beaucoup d'autres especes. Le Roy s'en est beaucoup amusé."

28 For the ostrich egg, see R to Mazzoleni, 30 August 1748, Biblioteca Laurenziana, Ms. Ashburnham 1522. Michel Adanson, a young protégé of Réaumur, left for Senegal in 1749 with instructions to observe ostriches in the wild and to ship eggs home if possible. *Art de faire*, 2:216. Réaumur's book also included his tricks for egg preservation, so that travelers might send home fertile eggs of exotic species to be hatched in Paris.

29 The Duc d'Orléans had wanted to replicate the Egyptian ovens in Paris during his regency. Réaumur had solicited a report of local village practices from the French consul in Cairo at that time, and based the account of the Egyptian ovens on that report. "We would have had the pleasure of seeing in Paris several thousand chicks born on a single day and in an oven similar to those of Egypt, if death had not taken that prince [the regent] from us too soon." *Art de faire*, 1:5–6.

30 Ibid., 1:30.

31 Languet de Gergy had been an active honorary member of the Société des Arts in the 1730s. He invented a method of spinning and weaving mousseline, for which he had a monopoly; women produced this fabric in the workshops of the community. On Languet, see Hamel, *Histoire de l'église*, 159–96.

32 *Art de faire*, 1:39.

33 Ibid., 1:40.

34 Ibid., 1:44.

35 On the nun known as Sister Marie, *Art de faire*, 1:285, 2:50–51.

36 "Notre curé de St. Sulpice qui scait habilement et en grand tirer parti de tout a une communauté de filles appellée l'enfant jesus, tout pres de Paris, dont il est le fondateur; il y a fait un etablissment pour faire eclore les poulets par l'une et par l'autre de mes manieres, qui a tres bien reussi; les poulets y fourmillent de toutes parts quoiqu'il en vende journellement." R to Mazzoleni, 30 August 1748, Biblioteca laurenziana, Ms. Ashburnham 1522.

37 For R's friendship with Mme de Rossignol, abbess of Malnoue, see *Art de faire*, 1:58. Marie-Anne Dumoustier, another sister, had died in 1739; Hélène Catherine lived in the convent until her death in 1761. Mme. de Rossignol had wel-

comed Réaumur and Dumoustier as visitors to the convent when it was still in Malnoue, a few hours ride from Paris.

38 Ibid., 1:57–63.

39 "Ces curiosités sont aussi offertes au public qui désire en avoir quelque connoissance." Blondel, *Architecture françoise*, 2:135.

40 R to Lord Morton, 5 May 1749, Beinecke Library, OSB Mss. 72.

41 For the different species in the yard, see *Art de faire*, 2:147; on the sheep, R to Mazzoleni, 21 May 1746, Biblioteca laurenziana, Ms. Ashburnham 1522. Guettard mentions the deformed sheep in his inventory of the collection, BCMHN, 1929/iii.

42 R to Lord Morton, 5 May 1749, Beinecke Library, OSB Mss, 72: "I did not think I should give in to entreaties to publish it sooner. I was quite happy to repeat my experiments in the most troublesome seasons, in winter and spring." He wrote in a similar vein to Mazzoleni, 30 August 1748, Bibliotheca laurenziana, Ms. Ashburnham 1522.

43 R, *Pratique de l'art de faire éclorre et d'élever en toute saison des oiseaux domestiques* (Paris: Imprimerie royale, 1751).

44 *Art de faire*, 1:18.

45 See R to Trembley, 16 March 1750, *Corr. inédite*, 336–37. The English translation was published as *The Art of Hatching and Bringing Up Domestick Fowls of All Kinds, at Any Time of the Year* (London: C. Davis, 1750). The book was also translated into Dutch, and C. Wolf published a summary in German. Trembley's extract was presented to the Royal Society and printed separately in London and Dublin.

46 R to Trembley, 26 May 1750, *Corr. inédite*, 339–41. Trembley's letters from this period are missing. Réaumur continued to give advice about thermometers in R to Trembley 27 June 1750 and 26 July 1750, *Corr. inédite*, 343–47. On the problems of consistency in Réaumur thermometers, see Gauvin, "Instrument That Never Was."

47 R to Bonnet, 18 December 1749. BG, Ms. Bonnet 24.

48 Bonnet to R, 9 August 1751, BG, Ms. Bonnet 70 (Bonnet's copy; original missing from Paris AAS).

49 Ludot to R, 24 December 1749, AAS-FR, Correspondance.

50 On Ludot, see Socard, *Biographie des personnages de Troyes*, 275. He was trained as a lawyer, but never practiced. He sometimes toyed with the idea of seeking a situation in the capital where he could indulge his avocation for natural history more fully. See R to Ludot, 8 November 1752, BL, Add. Ms. 39757.

51 R to Morton, 23 January 1750. Beinecke Library, OSB Mss. 72. Brisson made thermometers for distribution to Réaumur's acquaintances, but this was not a business. There were artisan glassmakers in the neighborhood who started

producing thermometers to Réaumur's specifications for their customers; Nollet also sold thermometers from his instrument shop. See R to Trembley, 26 May and 27 June 1750, *Corr. Inédite*, 339–40, 344.

52 R to Artur, 18 February 1752, in Chaia, "Sur une correspondance," 133.

53 *Art de faire*, 1:xi. This was the beginning of a period of "agromania," evident in the explosion of books, starting in the 1750s, on agricultural improvement and related topics. Shovlin, *Political Economy of Virtue*, 51–56.

54 Baux to R, 20 July 1751, AAS, dossier Baux.

55 Laurencys to R, 8 August 1756, AS-FR Correspondance. On verifying thermometers in his ovens before sending them out, R to Trembley, 26 July 1750, *Corr. inédite*, 346.

56 Faget de Saint Martial to R, 28 March 1755, AAS-FR, Correspondance.

57 Picault de la Rimbertière to R, 19 December 1749, AAS-FR, Correspondance.

58 Both citations from Le Duc to R, 26 March 1750, AAS-FR, Correspondance.

59 Le Duc to R, 26 March 1750 and 30 May 1750, AAS-FR, Correspondance.

60 Mugnier to R, 18 April 1754, AAS-FR, Correspondance. Body heat on Réaumur's thermometric scale is approximately 32 degrees, roughly the same as the temperature under a setting hen.

61 Ibid.

62 Mugnier to R, 19 July1754, AAS-FR, Correspondance.

63 "Mes termometres sont des votres, car ils ont été pris chez vous." Comte de Rochemore (from Le Grand Galargue en Languedoc) to R, 5 November 1756, AAS-FR, Correspondance. For another expensive chicken oven, see the discussion of Lord Stanhope's enormous construction: R to Trembley, 26 May 1750, *Corr. inédite*, 339.

64 Bonnel to R, 7 January 1753, AAS-FR, Correspondance.

65 Bonnel to R, 13 January 1753, AAS-FR, Correspondance.

66 Bourgine to R, 24 August 1751, AAS-FR, Correspondance. Bourgine signed himself "juge de la monnoie," La Rochelle. In his next letter, Bourgine described the two-level structure housing his ovens and artificial mothers (for raising chicks). The building was ninety-six feet long by eighteen wide, with the incubators below the nurseries. Bourgine to R, 2 November 1751, AAS-FR, Correspondance. Bourgine apparently belonged to a prominent merchant family in La Rochelle; he spent the last part of his life in India.

67 "Si je me suis determiné à le faire ce n'est que par les instances que me fait pour ainsi dire toutte la ville de faire un etablissement dans une province ou la volaille est tres rare et dont les armaments sur mer font une grande consommation, mais comment aider et eclaircir ceux qui voudront l'entreprendre si je suis moi meme un aveugle?" Bourgine to R, 24 August 1751, AAS-FR, Correspondance.

68 Bourgine to R, 2 November 1751, AAS-FR, Correspondance. From this point on, Bourgine's letters include greetings for Brisson and Mlle. Dumoustier.

69 Lanux thanks R for sending thermometers and reports on his chicken ovens; he

also tried incubation using "the method of the Chinese." Lanux to R, 7 September 1753, AAS, dossier Lanux. Bourgine refers to Lanux as "un curieux zélé." Bourgine to R, 16 September 1753.

70 Bourgine (from Chandernagor) to R, January 1754: "Je vous supplie toujours de me continuer votre protection auprès de messieurs de la compagnie pour mon avancement."

71 Boxes sent to Réaumur via the Compagnie des Indes and received in Lorient could well have been sent by Bourgine. "Notes des articles destinés pour M. de Réaumur, expediés à l'adresse de la Comp.ie des Indes le 11 8bre 1754," cited in McClellan and Regourd, *Colonial Machine*, 143n457.

72 On Gaultier's assistance to Pehr Kalm, the Linnaean botanist, see Müller-Wille, "Walnuts at Hudson's Bay." Gaultier's predecessor as king's physician in Quebec, Michel Sarrasin (sometimes spelled Sarrazin), had also served as correspondent of the Academy of Sciences. On Sarrasin, see Torlais, *Réaumur*, 282–86.

73 Gaultier to R, 30 October 1750, AAS, dossier Gaultier. The letter also includes a detailed account of muskrat behavior.

74 Gaultier called the nuns' operation "*une manufacture de poulets.*" Gaultier to R, 28 October 1753, AAS, dossier Gaultier.

75 Both quotations from Gaultier to R, 3 November 1753, AAS, dossier Gaultier.

76 For the vulture, see R, "Sur la digestion des oiseaux. Second mémoire: De la manière dont elle se fait dans l'estomac des oiseaux de proie," *MAS* (1752), 461-95. For the stork, AAS-FR, dossier 20, draft ms.: "Mémoire sur les aliments des oiseaux."Guettard's inventory of the collection includes hens sent by Bazin and "a black stork that lived in the courtyard of the house," where he had no doubt encountered it when it was still alive. Guettard, "Inventaire du cabinet de M. de Réaumur," BCMHN 1929 (iii), ff. 122v (hens), 103v (stork).

77 R to Mazzoleni, 26 August 1745, Biblioteca laurenziana, Ms. Ashburnham 1522.

78 *Art de faire*, 2:310–22, journal mentioned on 318. The journal is not in the archives.

79 Ibid., 2:322.

80 R to Mazzoleni , 26 August 1745, Biblioteca laurenziana, Ms. Ashburnham, 1522. On Diderot's reading of the rabbit-hen episode, Mortier, "Note sur un passage." Comte de Rochemore was inspired by the story to attempt his own interspecies crosses: "I have mated [*accouplé*] in one coop a rabbit, a duck and a hen; I will see if that produces anything." Rochemore to R, 5 November 1756 AAS-FR, Correspondance.

81 *Art de faire*, 2:302.

82 Ibid., 2:292.

83 Ms. notes on chickens, AAS-FR, dossier 25.

84 *Art de faire*, vol. 2, quotations on 228, 230.

85 Réaumur was referring to Maupertuis's conjectures about generation in *Vénus physique* (1745). These comments could equally well apply to Buffon's theory

in volume 2 of *Histoire naturelle* (Buffon, *Oeuvres complètes*, vol. 2). Although that volume was printed in September 1748, it was only released when all three volumes were completed, in September 1749. (*Art de faire* was already printed and bound by August 1749; see R to Trembley, 19 August 1749, *Corr. inédite*, 324.) It is not clear whether Réaumur had seen Buffon's text when he wrote this passage; rumors about Buffon's theory may well have been circulating at this time. See Terrall, "Speculation and Experiment." On the delays in printing and publication of Buffon's book, see Weil, "Correspondance Buffon-Cramer."

86 In one note he says the egg in the female provides "le terrain qui ne fournit qu'à la végétation d'un petit être organisé." Autograph ms. notes (1747), AAS-FR, dossier 25.

87 On Maupertuis's breeding experiments, see Terrall, *Man Who Flattened the Earth*, 334–40. Maupertuis had been a friend of Réaumur's in the 1730s and had contributed specimens to his collection, but they had fallen out by the 1740s and do not seem to have communicated directly about these experiments.

88 All quotations from manuscript notes in AAS-FR, dossier 25 ("Couvées des poules").

89 Ibid.: "Le 18 juin il m'est né dans ma couche un poulet de la poule à 5 doigts de cette loge qui déroute absolument tout systeme d'embryon du en entier soit au masle soit à la femelle."

90 Ibid.: "Quoiqu'il en soit ce poulet tient du masle par une de ses pattes & de la femelle par l'autre. Ils ont donc contribué l'un et l'autre à la production du germe. Voila une consequence bien embarassante."

91 Ibid.: "Le poulet avoit peri dans la coquille mais ce qui déroute absolument toutes mes observations precedentes c'est que le poulet avoit 5 doigts bien formés, bien distincts, à chaque patte. Le poulet avoit donc le nombre des doigts de la mere & non celui des doigts du pere. Cette poule n'a eu aucune communication avec un coq à 5 doigts. Que conclure de cette experience?"

92 *Art de faire*, 2:333–35.

93 Maupertuis, "Sur la génération des animaux" (1752), in *Oeuvres*, 2:307.

94 Pignon to R, 28 May 1756, AS-FR, Correspondance. Réaumur mentions Pignon's occupation in R to Séguier, 5 May 1749, in *Lettres inédites*, 76. Pignon also sent items for his collection, e.g., "a chest filled with asbestos and curious fossils." R to Mazzoleni, 3 September 1751, Biblioteca laurenziana Ms. Ashburnham 1522.

95 Antelmi to R, 12 February 1752, AAS-FR, Correspondance.

96 Ibid.

97 Antelmi, like many correspondents writing for the first time, also offered to send birds for Réaumur's collection.

98 See Buffon's discussion of his microscopical observations, with illustrations, in *Histoire naturelle*, vol. 2, chaps. 6 and 8. Buffon, *Oeuvres complètes*, 2: 244–92, 317–55.

99 The female fluids were taken from "glandular bodies" on the ovaries, or Graafian follicles, which Buffon considered analogous to the male testicles. On Buffon's observations, Marc Ratcliff, *Quest for the Invisible*; Mazzolini and Roe, "Introduction," in *Science against the Unbelievers*.

100 AAS p-v, 14 December 1748. The published version of this paper did not appear until 1752: Buffon, "Découverte." The plates had already appeared in the second volume of Buffon's *Histoire naturelle* in 1749. In May 1748, Buffon deposited with the secretary of the academy a sealed memorandum (*pli cacheté*) outlining his theory of generation, a common strategy for preserving priority; this document was not opened until 1866. The text is reproduced as Annexe 1, "Pli cacheté, déposé à l'Académie des Sciences le 17 mai 1748," in Buffon, *Oeuvres complètes*, 2:611–15.

101 Buffon to Cramer, 14 December 1748, in Weil, "Correspondance Buffon-Cramer," 132.

102 On the publication history of the *Histoire naturelle*, see Schmitt, "Introduction," in Buffon, *Oeuvres complétes*, vol. 2; Roger, *Buffon*.

103 R to Seguier, 25 May 1749, in *Lettres inédites*, 78–79.

104 The prospectus for *Histoire naturelle* assigned ornithology to the ninth volume, admitting that the bird collection was not yet "complete." "Programme de l'Histoire naturelle," in Buffon, *Oeuvres complètes* 1:843–49. Réaumur never completed his work on ornithology, though many chapters survive in manuscript. "Mémoires sur l'ornithologie," AAS-FR, dossier 22.

105 See, for example, Malesherbes, *Observations sur l'histoire naturelle*.

106 Hoquet, *Buffon*. As the editors of the modern edition of the *Histoire naturelle* note, Buffon's anthropocentric scheme reprises that of Aristotle. On this point, Buffon, *Oeuvres complètes*, 1:180n119, 183n127.

107 R, "Cabinets," undated autograph ms. fragment [c. 1750], AAS-FR, carton 1: "Je ne dirai rien contre une methode qui a revolté tous ceux qui étudient l'histoire naturelle. Il n'y a pas a craindre malgré la maniere dont elle est exposée qu'elle séduise ceux qui commenceront a s'appliquer a cette science, ils voudront des caracteres surs, invariables qui leur apprennent la place dans laquelle chaque etre doit etre placé. Et si cet ordre comme le veut cette methode, étoit reglé par le rapport qu'ils ont avec nous, ils se trouveroient arranger d'une maniere par trop bizarre. Nous ne scavons pas encore si l'autheur placeroit bien loin les animaux qui nous sont nuisibles de ceux qui nous sont utiles. Mais on voit au moins que les chiens et les chats sauvages ne seront pas placer dans la mesme classe que les chiens et les chats domestiques, que l'animal qui est le plus utile aux habitants d'un pays, et qui par consequence merite dans ce pays-la le 1er. rang, sera rejetter tres loin dans un autre pays. Le naturaliste du Canada placera au premier rang le castor. A la Chine, le cochon sera placé avant les boeufs. Dans chaque partie de l'Europe, dans chaque contrée de ces parties, il faudra suivre des ordres differents, &c."

108 On Buffon's style in the context of contemporary natural history works, see Loveland, *Rhetoric and Natural History*, chap. 1.

109 On the reception of Buffon's first volumes, see Hoquet, *Buffon*, 51–64. On Buffon's lack of credibility as a naturalist, see Malesherbes's assessment: "En effet, il est certain que ceux pour qui M. de Buffon a écrit ne sont point les naturalistes; je suis même persuadé que quand il a entrepris son ouvrage, il les connoissoit à peine de nom." Malesherbes, *Observations*, 204.

110 Buffon, *Histoire naturelle*, 2:255.

111 Roe, "Buffon and Needham"; Roe, "John Turberville Needham."

112 Buffon, *Histoire naturelle*, in *Oeuvres complètes* 2:292.

113 Needham was best known for his work on organic infusions. Watching an infusion of wheat, for example, he saw the fibers "swell from an interior Force so active, and so productive, that even before they resolved into, or shed any moving Globules, they were perfect Zoophytes teeming with life, and self-moving." Needham, "Summary," 645.

114 Trembley to Bentinck, 9/20 January 1750. BL Mss Egerton 1726; cited in *Corr. inédite*, 330.

115 See R to Trembley, 8 November 1749, *Corr. inédite*, 326, for observations of polyps in Poitou by Réaumur, Brisson, Lignac, and Dumoustier. Lignac's letters to Réaumur from the 1730s include his observations of insects; he later wrote a study of aquatic spiders (1749).

116 R to Mazzoleni, 11 December 1751, Biblioteca Laurenziana, Ms. Ashburnham 1522. He told Bonnet about their observations in similar terms; R to Bonnet, 10 December 1751, BG, Ms Bonnet 26.

117 Albrecht von Haller also challenged Buffon's observational acumen, especially with respect to his claims about the female reproductive glands and fluids. Haller's critique appeared as the preface to the German translation of Buffon and was reprinted in French translation. Haller, *Réflexions sur le système*.

118 R to Bonnet, 10 December 1751, BG, Ms. Bonnet 26.

119 Lignac, *Lettres à un américain*. One extant copy has a manuscript note on the flyleaf, saying it was printed at the Arsenal, "chez Mme. la duchesse du Maine, protectrice de Réaumur," the implication being that Réaumur arranged for the printing, though there is no way to assess the credibility of this source. (See Roger, *Sciences de la vie*, 692n60.) Réaumur sent Lignac's book to Bonnet, Trembley, Mazzoleni, Séguier, Cramer, and no doubt many others. On Lignac's book, and especially its theological and metaphysical commitments, see Roger, *Sciences de la vie*, 691–703. For the critiques of Buffon, see Hoquet, *Buffon*, 51–64.

120 "Quelque fois un peut trop teologique, le stil en est clair, agreable & assez precis. Le palais de verre ou logoit l'Academicien est réduit en poudre." Bonnet to Roger, 15 December 1752, BG, Ms. Bonnet 70.

121 Réaumur's notes on the infusion observations: "Animalcules des infusions," holograph ms., AAS-FR, dossier 27.

122 John Hill attempted a description of many kinds of microorganism in 1752; Spallanzani took this much further in the 1780s. On this and subsequent decades of microscopical research, see Ratcliff, *Quest for the Invisible*, chap. 8.

Epilogue

1 "Somme des depenses qu'exigera chaqu'année le cabinet de Mr. de Réaumur" [n.d.], (autograph ms.), AAS, dossier Réaumur.

2 On this project, and especially its completion by Duhamel de Monceau, see Gillispie, *Science and Polity*, 337–56.

3 The works on ants and beetles were eventually published nearly two hundred years after Réaumur's death. Caullery, *Papiers laissés par Réaumur*; R, *Natural History of Ants.*

4 For the disposition of Réaumur's estate, see Torlais, *Réaumur*, 380–88.

5 Brisson based his *Règne animal*, of which only the quadrupeds volume was published, on the specimens in Réaumur's collection; it can thus be read as a partial catalog of the quadrupeds.

6 Buffon, *Histoire naturelle, générale et particulière*, vols. 16–24 (Paris: Imprimerie Royale, 1770–1783). Illustrations were by Martinet, who had made the drawings for Brisson's book.

7 See Llana, "Natural History." Another element in this story is the cooptation by Diderot of some of the engraved plates prepared for the *Description des arts et métiers* under Réaumur's supervision. Gillispie, *Science and Polity.*

8 Maupertuis, *Lettre sur le progrès des sciences* (1756), in *Oeuvres*, 2:418.

9 Diderot, *Pensées sur l'interprétation de la nature*, paragraph LIV, 86.

10 Ibid., XII, 36–37.

11 Nollet, *Discours*, 24–25.

12 See Gillispie, "Natural History of Industry."

13 Hoquet, "History without Time." See also Hoquet, *Buffon*, chap. 1, "L'*Histoire naturelle* n'est pas une histoire naturelle."

14 Singy, "Huber's Eyes."

15 Sleigh, *Six Legs Better.*

16 The neuroethologist is Jochen Smolka; his work, published in *Current Biology*, was described in "Beetles' Affinity for Dung May Keep Them Cool," *New York Times*, 30 October 2012, D3.

17 A striking example is Thomas Eisner's book *For Love of Insects*, which incorporates the same kind of discovery narrative used by Réaumur to chronicle investigations of the biologically active chemicals used by insects in myriad ways.

BIBLIOGRAPHY

Full references to manuscript sources and eighteenth-century journal articles are given in the notes.

Manuscript Collections

Archives nationales de France, Paris
Archives de l'Académie des Sciences, Paris
Beinecke Library, Yale University
Bibliothèque de Genève, Geneva
Bibliothèque cantonale et universitaire, Lausanne
Bibliothèque centrale du Muséum d'histoire naturelle, Paris
Bibliothèque centrale, Université de Mons
Biblioteca laurenziana, Florence
Bibliothèque municipale de Nîmes
Bibliothèque municipale de La Rochelle
Bibliothèque publique et universitaire, Neufchâtel
British Library, London
Getty Research Institute Library, Los Angeles
Houghton Library, Harvard University

Primary Sources

Adanson, Michel. *Histoire naturelle du Sénégal: Coquillages, avec la relation abrégée d'un voyage fait en ce pays, pendant les années 1749, 50, 51, 52 & 53*. Paris: C.-J.-B. Bauche, 1757.

Bazin, Gilles-Augustin. *Abrégé de l'histoire naturelle des insectes, pour servir de suite à l'histoire naturelle des abeilles*. 4 vols. Paris: Les frères Guerin: 1747–1751.

———. *Histoire naturelle des abeilles*. Paris: Les frères Guerin, 1744.

Blondel, Jacques-François. *L'architecture françoise, ou Receuil des plans, élévations, coupes et profils des églises, maisons royales, palais, hôtels & edifices les plus considérables de Paris*. 4 vols. Paris: C. A. Jombert, 1752–1756.

Bonnet, Charles. *Mémoires autobiographiques*. Edited by Raymond Savioz. Paris: J. Vrin, 1948.

———. *Traité d'insectologie, ou Observations sur les pucerons* (1st ed., Paris: Durand, 1745). Vol. 1 of *Oeuvres d'histoire naturelle et de philosophie de Charles Bonnet*. 18 vols. Neuchâtel: Samuel Fauche, 1779–1783.

Brisson, Mathurin-Jacques. *Ornithologie, ou méthode contenant la division des oiseaux en ordres, sections, genres, especes & leurs variétés*. 6 vols. Paris: Bauche, 1760.

———. *Le règne animal divisé en IX classes, ou méthode contenant la division générale des animaux*. Paris: Bauche, 1756.

Buffon, Georges-Louis Leclerc de. "Découverte de la liqueur séminale dans les femelles vivipares et du réservoir qui la contient." *MAS* (1748), 211–28.

———. *Oeuvres complètes*. Vols. 1–3. Edited by Stéphane Schmitt. Paris: Honoré Champion, 2007–2009.

Caullery, Maurice, ed. *Les papiers laissés par de Réaumur et le Tome VII des Mémoires pour servir à l'histoire des insectes.* Paris: Paul Lechevalier, 1929.

Chaia, Jean. "Sur une correspondance inédite de Réaumur avec Artur, premier médecin du Roy à Cayenne." *Episteme* 2 (1968): 36–57, 121–38.

Cordier, Samuel, and François Pugnière, eds. *Jean-François Séguier, Pierre Baux: Lettres, 1733–1756*. Avignon: Editions A. Barthélemy, 2006.

Daubenton, Louis-Jean-Marie. "Description du cabinet du Roy." In Buffon, *Oeuvres complètes*, 3:89–98.

Demeulenaere-Douyère, Christine, and David Sturdy, eds. *L'enquête du Régent, 1716–1718: Sciences, techniques et politique dans la France pré-industrielle.* Turnhout, Belgium: Brepols, 2008.

Dézallier d'Argenville, Antoine-Joseph. *Histoire naturelle éclaircie dans deux de ses parties principales, la lithologie et la conchyologie.* Paris: De Bure, 1742. Rev. ed. 1757.

Diderot, Denis. *Pensées sur l'interpretation de la nature*. Edited by Jean Varloot. *Oeuvres complètes*, vol. 9. Paris: Hermann, 1981.

Geoffroy, Claude-Joseph. "Observations sur les vessies qui viennent aux ormes, & sur une sorte d'excroissance à peu-près pareille qui nous est apportée de la Chine." *MAS* (1724), 320–26.

Gersaint, E. F. *Catalogue raisonné de coquilles, insectes, plantes marines, et autres curiosités naturelles.* Paris: Prault fils, 1736.

Guettard, Jean-Etienne. *Mémoires sur différentes parties des sciences et des arts.* 5 vols. Paris: Prault, 1768–1783.

Haller, Albrecht von. *Réflexions sur le système de la génération de M. de Buffon*. Geneva: Barillot et fils, 1751.

Lesser, Friedrich Christian. *Théologie des insectes ou démonstration des perfections de Dieu dans tout ce qui concerne les insectes . . . avec des remarques de M. P. Lyonnet* [sic]. 2 vols. La Haye: Jean Swart, 1742; Paris: L. Durand, 1745.

Lignac, Joseph-Adrien Lelarge de. *Lettres à un américain sur l'histoire naturelle, générale et particulière de Monsieur de Buffon.* 5 vols. Hambourg [Paris], 1751.

Lyonet, Pierre. Notes to Lesser, *Théologie des insectes*.

Maupertuis, P.-L. M. de. *Vénus physique*. N.p, 1745.

———. *Oeuvres de M. de Maupertuis*. 4 vols. Lyon: Bruyset, 1768.

Musset, Georges, ed. *Lettres inédites de Réaumur.* La Rochelle: Mareschal & Martin, 1886.

Needham, John Turberville. *Nouvelles découvertes faites avec le microscope . . . avec un mémoire sur les polypes à bouquet et sur ceux en entonnoir par A. Trembley*. Leiden: E. Luzac, 1747.

———. "A Summary of Some Late Observations upon the Generation, Composition, and Decomposition of Animal and Vegetable Substances." *Philosophical Transactions* 45 (1748): 615-66.

Nollet, Jean-Antoine. *L'art des expériences ou avis aux amateurs de la physique sur le choix, la construction et l'usage des instrumens.* 3 vols. Paris: Durand, 1770.

———. *Discours sur les dispositions et sur les qualités qu'il faut avoir pour faire du progrès dans l'etude de la physique expérimentale.* Paris, 1753.

———. *Leçons de physique expérimentale*. 6 vols. Paris: Les frères Guérin, 1743–1748.

———. *Programme, ou idée générale d'un cours de physique expérimentale.* Paris: Les frères Guerin, 1738.

Réaumur, R. A. F. de. *L'art de convertir le fer forgé en acier, et l'art d'adoucir le fer fondu, ou de faire des ouvrages de fer fondu, aussi finis que de fer forgé*. Paris: Michel Brunet, 1722.

———. *L'art de faire éclorre et d'éléver en toute saison des oiseaux domestiques de toutes espèces*. 2 vols. Paris: Imprimerie royale, 1749. 2nd ed., 1751.

———. "Histoire des teignes ou des insectes qui rongent les laines et les pelleteries: Premiere partie." *MAS* (1728), 139–58.

———. *Mémoires pour servir à l'histoire des insectes.* 6 vols. Paris: Imprimerie royale, 1734–1742.

———. *The Natural History of Ants, from an Unpublished Manuscript in the Archives of the Academy of Sciences of Paris.* Bilingual edition. Translated and with notes by William Morton Wheeler. New York: Alfred A. Knopf, 1926.

———. "Suite de l'histoire des teignes ou des insectes qui rongent les laines et les pelleteries: Seconde partie." *MAS* (1728), 311–37.

[Réaumur, R. A. F. de.] *Différens moyens d'empêcher de se corrompre les oiseaux morts qu'on veut envoyer dans les pays éloignez, & de les y faire arriver bien conditionnez.* Paris: Imprimerie royale, n.d. [1745].

Senebier, Jean. *L'art d'observer*. Geneva: Philibert & Chirol,1775.

Smith, Cyril Stanley, ed. *Réaumur's Memoirs on Steel and Iron*. Translated by A. G. Sisco. Chicago: University of Chicago Press, 1956. (Translation of *L'art de convertir le fer forgé en acier.*)

Spallanzani, Lazzaro. *Carteggi di Lazzaro Spallanzani*. 12 vols. Edited by Pericle Di Pietro. Modena: Mucchi, 1984–1990.

———. *Expériences pour servir à l'histoire de la génération des animaux et des plantes*. Translated by J. Senebier. Geneva: Chirol, 1785.

Swammerdam, Jan. *Histoire générale des insectes*. Utrecht: Guillaume de Walcheren, 1682.

Tafel, R. L., ed. *Documents concerning the life and character of Emanuel Swedenborg*. 3 vols. London: Swedenborg Society, 1875.

Trembley, Abraham. *Mémoires pour servir à l'histoire des polypes d'eau douce*. Leiden: Jean & Herman Verbeek, 1744.

Trembley, Maurice, ed. *Correspondance inédite entre Réaumur et Abraham Trembley*. Geneva: Georg & Compagnie, 1943.

Weil, Françoise. "La correspondance Buffon-Cramer," *Revue d'histoire des sciences* 14 (1961): 97–136.

Secondary Sources

Algazi, Gadi. "Scholars in Households: Refiguring the Learned Habitus, 1480–1550." *Science in Context* 16 (2003): 9-42.

Audisio, Gabriel, and François Pugnière. *Jean-François Séguier: Un Nîmois dans l'Europe des Lumières*. Aix-en-Provence: Edisud, 2005.

Babelon, Jean-Pierre. "L'Hôtel de Rambouillet." *Paris et Ile-de- France: Mémoires publiés par la Féderation des Sociétés historiques et archéologiques de Paris et de L'Ile-de-France* 11 (1960): 313–49.

Badinter, Elisabeth. *Les passions intellectuels: Desirs de gloire (1735–1751)*. Paris: Fayard, 1999.

Bell, David. *Lawyers and Citizens: The Making of a Political Elite in Old Regime France*. New York: Oxford University Press, 1994.

Bertoloni Meli, Domenico. *Mechanism, Experiment, Disease: Marcello Malpighi and Seventeenth-Century Anatomy*. Baltimore: Johns Hopkins University Press, 2011.

Birembaut, Arthur. "La contribution de Réaumur à la thermométrie." In Grassé, *Vie et l'oeuvre*.

Bleichmar, Daniela. "Learning to Look: Visual Expertise across Art and Science in Eighteenth-Century France." *Eighteenth-Century Studies* 46 (2012): 85–112.

Bluche, François. "Les officiers du grenier à sel de Paris au XVIIIe siècle." *Paris et Ile-de-France* 21 (1966): 293–336.

Bontemps, Daniel, and Catherine Prade. "Un magasin parisien d'ouvrages en fonte de fer ornée au XVIIIe siècle: Une réussite méconnue de Réaumur." *Bulletin de la Société de l'histoire de Paris et de l'Ile-de-France* 118 (1991): 215–61.

Bourguet, Marie-Noëlle. "Measureable Difference: Botany, Climate, and the Gardener's Thermometer in Eighteenth-Century France." In *Colonial Botany: Science, Commerce, and Politics in the Early Modern World*, edited by Londa

Schiebinger and Claudia Swan, 270–86. Philadelphia: University of Pennsylvania Press, 2005.

Chapron, Emmanuelle. "Les échanges savants à l'épreuve de la distance: Jean-François Séguier (1703–1784) entre Vérone et Nîmes." *Rives méditerranéennes* (2009). http://rives.revues.org/2954 (consulted 13 March 2010).

Choudhury, Mita. *Convents and Nuns in Eighteenth-Century French Politics and Culture*. Ithaca, NY: Cornell University Press, 2004.

Clark, William. "The Death of Metaphysics in Enlightened Prussia." In *The Sciences in Enlightened Europe*, edited by W. Clark, J. Golinski, and S. Schaffer, 433–73. Chicago: University of Chicago Press, 1999.

Cobb, Matthew. "Malpighi, Swammerdam and the Colourful Silkworm: Replication and Visual Representation in Early Modern Science." *Annals of Science* 59 (2002): 11–147.

Cook, Harold J. "Time's Bodies: Crafting the Preparation and Preservation of Naturalia." In *Merchants and Marvels: Commerce, Science and Art in Early Modern Europe*, edited by P. H. Smith and P. Findlen. New York: Routledge, 2002.

Cooper, Alix. "Homes and Households." In *The Cambridge History of Science*, vol. 3, *Early Modern Science*, edited by Katharine Park and Lorraine Daston, 224–37. Cambridge: Cambridge University Press, 2006.

Daston, Lorraine. "Attention and the Values of Nature in the Enlightenment." In *The Moral Authority of Nature*, edited by L. Daston and F. Vidal, 100–126. Chicago: University of Chicago Press, 2004.

Daston, Lorraine, and Peter Galison. *Objectivity*. New York: Zone Books, 2007.

Daston, Lorraine, and Elizabeth Lunbeck, eds. *Histories of Scientific Observation*. Chicago: University of Chicago Press, 2011.

Dawson, Virginia P. *Nature's Enigma: The Problem of the Polyp in the Letters of Bonnet, Trembley and Réaumur*. Philadelphia: American Philosophical Society, 1987. (Appendix: Correspondence of Trembley and Bonnet.)

Dietz, Bettina. "Mobile Objects: The Space of Shells in Eighteenth-Century France." *British Journal for the History of Science* 39 (2006): 363–82.

Eisner, Thomas. *For Love of Insects*. Cambridge, MA: Harvard University Press, 2003.

Findlen, Paula. *Possessing Nature: Museums, Collecting and Scientific Culture in Early Modern Italy*. Berkeley: University of California Press, 1994.

———. "Science as a Career in Enlightenment Italy: The Strategies of Laura Bassi." *Isis* 84 (1993): 441–69.

———. "Science, History, and Erudition: Athanasius Kircher's Museum at the Collegio Romano." In *The Great Art of Knowing: The Baroque Encyclopedia of Athanasius Kircher*. Stanford: Stanford University Libraries, 2001.

Gauvin, Jean-François. "The Instrument That Never Was: Inventing, Manufacturing, and Branding Réaumur's Thermometer during the Enlighenment." *Annals of Science* 69 (2011): 514–49.

Gillispie, Charles Coulston. "The Natural History of Industry." In *Essays and*

Reviews in History and History of Science. Philadephia: American Philosophical Society, 2007.

———. *Science and Polity in France at the End of the Old Regime*. Princeton: Princeton University Press, 1980.

Glorieux, Guillaume. *A l'enseigne de Gersaint: Edme-François Gersaint, marchand d'art sur le pont Notre-Dame (1694–1750)*. Seyssel: Champ Vallon, 2002.

Golinski, Jan. *British Weather and the Climate of Enlightenment*. Chicago: University of Chicago Press, 2007.

Grassé, P., ed. *La vie et l'oeuvre de Réaumur (1683–1757)*. Paris: Presses universitaires de France, 1962.

Guerrini, Anita. "Duverney's Skeletons." *Isis* 94 (2003): 577–603.

Hamel, Charles. *Histoire de l'église de Saint Sulpice*. Paris: Lecoffre, 1900.

Hamonou-Mahieu, Aline. *Claude Aubriet: Artiste naturaliste des lumières*. Paris: CTHS, 2010.

Harkness, Deborah. *The Jewel House: Elizabethan London and the Scientific Revolution*. New Haven, CT: Yale University Press, 2007.

———. "Managing an Experimental Household: The Dees of Mortlake and the Practice of Natural Philosophy." *Isis* 88 (1997): 247–62.

Harth, Erica. *Cartesian Women: Versions and Subversions of Rational Discourse in the Old Regime*. Ithaca, NY: Cornell University Press, 1992.

Hoquet, Thierry. *Buffon: Histoire naturelle et philosophie*. Paris: Honoré Champion, 2005.

———. "History without Time: Buffon's Natural History as a Nonmathematical Physique." *Isis* 101 (2010): 30–61.

Hublard, Emile. *Le naturaliste hollandais Pierre Lyonet, sa vie et ses oeuvres d'après des documents inédites.* Brussels: J. Lebègue, 1910.

Hunter, Lynette, and Sarah Hutton. *Women, Science and Medicine, 1500–1700: Mothers and Sisters of the Royal Society.* Thrupp, Gloucestershire: Sutton, 1997.

Iliffe, Rob and Frances Willmoth. "Astronomy and the Domestic Sphere: Margaret Flamsteed and Caroline Herschel as Assistant-Astronomers." In Hunter and Hutton, *Women, Science and Medicine*.

Laissius, Yves. "Les cabinets d'histoire naturelle." In *Enseignement et diffusion des sciences en France au XVIII siècle*, edited by René Taton, 342–84. Paris: Hermann, 1964.

Licoppe, Christian. *La formation de la pratique scientifique: Le discours de l'expérience en France et en Angleterre, 1630–1820*. Paris: La Découverte, 1996.

Llana, James. "Natural History and the *Encyclopédie*." *Journal of the History of Biology* 33 (2000): 1–25.

Loveland, Jeff. *Rhetoric and Natural History: Buffon in Polemical and Literary Context*. Oxford: Voltaire Foundation, 2003.

Maistre, Myriam. *Les précieuses: Naissance des femmes de lettres en France au XVIIe siècle*. Paris: Champion, 1999.

Martin, Meredith. *Dairy Queens: The Politics of Pastoral Architecture from Catherine de Medici to Marie-Antoinette*. Cambridge, MA: Harvard University Press, 2011.

McClellan, James E., and François Regourd. *The Colonial Machine: French Science and Overseas Expansion in the Old Regime*. Turnhout, Belgium: Brepols, 2011.

McConnell, Anita. "The Flowers of Coral: Some Unpublished Conflicts from Montpellier and Paris during the Early 18th Century." *History and Philosophy of the Life Sciences* 12 (1990): 51–66.

Mazzolini, Renato, and Shirley Roe, eds. *Science against the Unbelievers: The Correspondence of Bonnet and Needham, 1760–1780*. Studies on Voltaire and the Eighteenth Century, vol. 243. Oxford: Voltaire Foundation, 1986.

Middleton, W. E. Knowles. *A History of the Thermometer and Its Use in Meteorology*. Baltimore: Johns Hopkins University Press, 1960.

Mortier, Roland. "Note sur un passage du *Rêve de d'Alembert*: Réaumur et le problème de l'hybridation." In Grassé, *Vie et l'oeuvre*, 116–23.

Müller-Wille, Staffan. "Walnuts at Hudson Bay." In *Colonial Botany: Science, Commerce, and Politics in the Early Modern World*, edited by Londa Schiebinger and Claudia Swan, 34–48. Philadelphia: University of Pennsylvania Press, 2005.

Ogilvie, Brian. *The Science of Describing: Natural History in Renaissance Europe*. Chicago: Chicago University Press, 2006.

Olivier, Marc. "Gilles Auguste Bazin's 'True Novel' of Natural History." *Eighteenth-Century Fiction* 18 (2005): 187–202.

Polderman, Marie. "Le manuscrit de Jacques François Artur: Appareil critique." In J. F. Artur, *Histoire des colonies françoises de la Guianne*. Paris: Ibis Rouge Editions, 2002.

Pomata, Gianna. "*Praxis historialis*: The Uses of '*Historia*' in Early Modern Medicine." In *Historia: Empiricism and Erudition in Early Modern Europe*, edited by Gianna Pomata and Nancy Siraisi. Cambridge, MA: MIT Press, 2005.

Pomian, Kristof. *Collectors and Curiosities: Paris and Venice, 1500–1800*. Cambridge: Polity Press, 1990.

Pyenson, Louis, and Jean-François Gauvin, eds. *The Art of Teaching Physics: The Eighteenth-Century Demonstration Apparatus of Jean-Antoine Nollet*. Quebec: Editions du Septentrion, 2002.

Ratcliff, Marc. *The Quest for the Invisible: Microscopy in the Enlightenment*. Burlington, VT: Ashgate, 2009.

———. "Trembley's Strategy of Generosity and the Scope of Celebrity in the Mid-Eighteenth Century." *Isis* 95 (2004): 555–75.

Roche, Daniel. *Les républicains des lettres: Gens de culture et lumières au XVIIIe siècle*. Paris: Fayard, 1988.

———. *Le siècle des lumières en province: Académies et académiciens provinciaux, 1680–1789*. Paris: Mouton, 1978.

Roe, Shirley. "Buffon and Needham." In *Buffon 88: Actes du colloque international pour le bicentenaire de la mort de Buffon*. Edited by J. Gayon and J.-C. Beaune, 439–50. Paris: J. Vrin, 1992.

———. "John Turberville Needham and the Generation of Living Organisms." *Isis* 74 (1984): 159–84.

Roger, Jacques. *Buffon: Un philosophe au Jardin du Roi*. Paris: Fayard, 1989.

———. *Les sciences de la vie dans la pensée française du XVIIIe siècle*. Paris: Armand Colin, 1971.

Ronfort, Jean-Nerée. "Science and Luxury: Two Acquisitions by the J. Paul Getty Museum." *J. Paul Getty Museum Journal* 17 (1989): 47–82.

Ronsseray, Céline. "Un destin guyanais: Jacques-François Artur, premier médicin du roi à Cayenne au XVIIIe siècle." *Annales de Normandie* 53 (2003): 351–80.

Ruestow, Edward. *The Microscope in the Dutch Republic: The Shaping of Discovery*. Cambridge: Cambridge University Press, 1996.

Shapin, Steven. "The House of Experiment in Seventeenth-Century England." *Isis* 79 (1988): 373–404.

Schiebinger, Londa. *The Mind Has No Sex? Women in the Origins of Modern Science*. Cambrige, MA: Harvard University Press, 1989.

Sheridan, Geraldine. "Recording Technology in France: The *Description des art*, Methodological Innovation and Lost Opportunities at the Turn of the Eighteenth Century." *Cultural and Social History* 5 (2008): 329–54.

Shovlin, John. *The Political Economy of Virtue: Luxury, Patriotism and the Origins of the French Revolution.* Ithaca, NY: Cornell University Press, 2006.

Singy, Patrick. "Huber's Eyes: The Art of Scientific Observation before the Emergence of Positivism." *Representations* 95 (2006): 54–75.

Sleigh, Charlotte. *Six Legs Better: A Cultural History of Myrmecology*. Baltimore: Johns Hopkins University Press, 2007.

Smith, Pamela. *The Body of the Artisan: Art and Experience in the Scientific Revolution*. Chicago: University of Chicago Press, 2004.

Socard, Emile. *Biographie des personnages de Troyes et du département de l'Aube*. Troyes: L. Lacroix, 1882.

Spary, E. C. "Codes of Passion: Natural History Specimens as a Polite Language in Late 18th-Century France." In *Wissenschaft als culturelle Praxis, 1750–1900*, edited by H. E. Bödeker, P. H. Reill, and J. Schlumbohm, 105–35. Göttingen: Vandenhoeck and Ruprecht, 1999.

———. "The Nature of 'Nature' in the Enlightenment." In *The Sciences in Enlightened Europe*, edited by W. Clark, J. Golinski, and S. Schaffer. Chicago: University of Chicago Press, 1999.

———. "Scientific Symmetries." *History of Science* 42 (2004): 1-46.

———. *Utopia's Garden: French Natural History from Old Regime to Revolution*. Chicago: University of Chicago Press, 2000.

Stalnaker, Joanna. *The Unfinished Enlightenment: Description in the Age of the Encyclopedia*. Ithaca, NY: Cornell University Press, 2010.

Sturdy, David. *Science and Social Status: The Members of the Académie des Sciences, 1666–1750*. Woodbridge, UK: Boydell Press, 1995.

Terrall, Mary. "Following Insects Around: Tools and Techniques of Natural History in the Eighteenth Century." *British Journal for the History of Science* 43 (2010), 573–88.

———. "Frogs on the Mantelpiece: The Practice of Observation in Daily Life." In Daston and Lunbeck, *Histories of Scientific Observation*.

———. *The Man Who Flattened the Earth: Maupertuis and the Sciences in the Enlightenment*. Chicago: University of Chicago Press, 2002.

———. "Speculation and Experiment in Enlightenment Life Sciences." In *Heredity Produced: At the Crossroads of Biology, Politics, and Culture, 1500–1800*, edited by Staffan Müller-Wille and Hans-Jörg Rheinberger, 253–75. Cambridge, MA: MIT Press, 2007.

Torlais, Jean. *L'abbé Nollet: Un physicien au siècle des lumières*. Paris: Sipuco, 1954.

———. *Réaumur: Un esprit encyclopédique en dehors de l'Encyclopédie*. Paris: Desclée de Brouwer, 1936. 2nd ed., Paris: Albert Blanchard, 1961.

Trinkle, Dennis. "Noel-Antoine Pluche's *Le spectacle de la nature*: An Encyclopedic Bestseller." *Studies on Voltaire and the Eighteenth Century* 358 (1997): 93–134.

Van Seters, W. H. *Pierre Lyonet, 1706–1789*. La Haye: Martinus Nijhoff, 1962.

Williams, Elizabeth A. *A Cultural History of Medical Vitalism in Enlightenment Montpellier*. Burlington, VT: Ashgate, 2003.

Wilson, Catherine. *The Invisible World: Early Modern Philosophy and the Invention of the Microscope*. Princeton, NJ: Princeton University Press, 1995.

INDEX